DRONE BOOK 2016

Contact Information:

Robokingdom LLC - United States of America

Email: editor@roboticmagazine.com

www.dronemall.com

www.robokingdom.com

About the Author: Since 2008, Robokingdom LLC runs several websites mostly on robotics subjects, including a news and information site on robots, roboticmagazine.com.

Publisher: Ahmet Tuter

Author: Robokingdom LLC

ISBN: 978-1-943605-00-2

Publication Date: June 20, 2016

Cover Photo Credits (in alphabetical order):

Alpha Unmanned Systems - Spain

www.alphaunmannedsystems.com

Altigator - Belgium

www.altigator.com

Flypro - China

www.flypro.com

Hubsan - China

www.hubsan.com

Iron Ridge - USA

www.ironridgeuas.com

Mavtech - Italy

www.mavtech.eu

Service Drone - Germany

www.service-drone.com

Traxxas - USA

www.traxxas.com

CONTENTS

Foreword

We operate a news website about robotics, RoboticMagazine.Com since 2008, which is about all types of robots and robotics subjects. This book came along as we saw the obvious increase in usage of drones.

This book is not only for hobbyists, whether beginner or experienced, but also for advanced users and companies, in other words, for anyone who is involved or interested in drones, in some way. Having operated a news site about all types of robots since many years, we did not want to limit the book to just one specific subject or audience.

The book has two main sections. The first part is mainly about important terms and concepts for drones, as well as a section that briefly gives information about making a hobby grade multirotor drone. Even when introducing terms and concepts, we wanted it to be more than a dull glossary with just definitions, but we also tried to give as much information as possible by photo examples from many companies, or various graphics. We also have sections where we describe basic steps to building a hobby grade multirotor drone, discuss important points when choosing parts, introduction to drone safety, drone photography and filming.

The second part of our book provides a list of sellers and manufacturers and their products, from all around the world. We contacted many companies while making this list, and for some, you can find more detailed information. The list contains manufacturers, parts manufacturers, for both hobby and advanced grade drones, sellers, wholesalers and service companies. After that, discussion forums and news websites list for drones is also included.

I hope you find this book helpful.

Sincerely,

Ahmet Tuter

Editor

RoboticMagazine.Com

Introduction

So, what are drones? They can be defined as robotic devices, without a human on board, that are either remote controlled by humans or moving autonomously. Almost everyone, immediately thinks that drones represent only flying unmanned vehicles, but technically it is not true. Unmanned Ground or Sea Vehicles are also drones. However, our focus in this book is flying drones, and from now when you see the word drone, please understand that we mean only flying unmanned robots. The main thing is that the drones are unmanned. Flying drones are also called UAV, which stands for Unmanned Aerial Vehicles or UA, Unmanned Aircraft, or UAS, Unmanned Aircraft System. Drone use mostly started more than 10 years ago in military applications at first, but especially as the battery and automatic pilot technologies improve and decrease in cost, we begin to see them increasingly in our daily lives, with longer flight times, faster speeds, higher payload capacities, more capable autonomous features.

Types of drones:

Planes: Unmanned planes follow the same principle as manned planes, which is based on lifting the aircraft by means of pressure differential lifting force at different sides of the wings, when a propeller in the front pushes the air towards the wings. Planes fly forward faster than helicopters or multirotor drones but they cannot hover in the air like them and do not have the ability to takeoff or land vertically.

Helicopters: Just like the real life helicopters, unmanned helicopters also have two rotors, one is the main bigger rotor that pushes the air downwards does the lifting, and also moves the helicopter forward and the other smaller propeller attached to the tail vertically for controlling which direction the helicopter goes. The small propeller at the tail also counteracts the rotation affect that the main propeller creates. Similar to multirotors, helicopters also use gyroscopes in order to balance themselves against the effects of wind, or when the helicopter throttles up and the turning effect increases.

Multi Rotors: This is the most popular type of drones in use today in general consumer and hobbyist market and we focus mostly on multirotors in this book. The multirotor system offers an important advantage in comparison to single rotor helicopters, which is the greater level of control and also the ability to stay in the air even if one of the rotors fail. More rotors of course enable greater carrying capacities, and mean smaller rotor blades which are easier to use. Of course a stronger and heavier battery is needed to run more blades. The most common type of multirotor drone is a quadcopter, which means, a drone with 4 blades. For heavier loads 6 (hexacopter) or even 8 (octacopter) propellers may be used.

Current Uses of drones

- Military
  Examples: Reconnaissance or Attack drones

- Visual inspection, photography or video recording
  Examples: Wildlife Control, Ski, Fire, Border Patrol, Stadium Recording, Surveying and Mapping, Traffic Control, Photo Taking, Film Recording, Construction or Existing Structures Inspection and many more...

- Delivery
  Example: Amazon is now seriously considering delivery by drones but trying to overcome obstacles about regulations. Drone delivery is expected to have a huge application in very near future

- Rescue and emergency deliveries
  Examples: Delivery of emergency supplies to mountaineers, rushing life vest to people in the sea, quickly searching an area for survivors after disasters

- Law enforcement
  Examples: Monitoring an area flexibly, safely and remotely, following of escaping suspects, safely delivering or removing something to / from a specific location

- Broadcasting
  Examples: Another internet giant, Facebook, is now considering beaming down internet through the use of drones.

Drone Components, Concepts and Systems:

Accelerometer: It is an electromechanical device that measures acceleration both horizontally in both horizontal axis due to speed, slow down, or high G turns, and vertically due to gravity and lift up or down. It also helps stabilizing the drone. Accelerometer and Gyroscope sensors are often combined in one sensor called Inertial Measurement unit and connected to flight controller. For photo and charts, see Inertial Navigation System, INS.

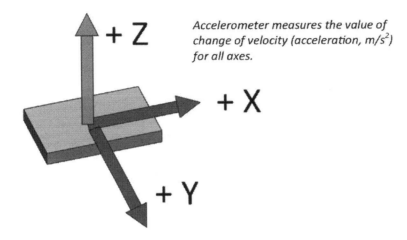

Accelerometer measures the value of change of velocity (acceleration, m/s²) for all axes.

Aileron: See Roll.

Airspeed Sensor: It measures the speed of the drone relative to the air, by measuring the positive and negative pressure differences around the drone. When purchased, they usually come together with pitot tube and connection cables. It is recommended for advanced users or drones only, as it necessitates an extra layer of control and tuning. Through pitot tube, the pressure is measured and then this is converted to air speed. Air speed varies with the square root of air pressure. The pitot tube, which takes in the air, transmits it to the sensor through rubber tubing. The sensor is connected to flight controller through a 4 wire I²C cable. Air speed of drone is different than its speed relative to ground. When calculating flight time for a certain distance, the ground speed is used. For example, if the aircraft is moving in the air with 200 km/h, into a headwind of 5 km/h, then its ground speed is 195 km/h. This is how fast the shadow of the aircraft moves on the ground. When airspeed is corrected for pressure and temperature, true airspeed is obtained. This is the true speed at which the aircraft moves through the air fluid that surrounds it.

Altimeter: This is for measuring the altitude of the drone, with respect to sea level. Altimeter can also be used to stabilize the height of the drone, when the operator sets it to hover mode.

Autopilot: See flight controller

Bank Angle: This is the angle between the longitudinal axis of the aircraft and the horizontal axis, when the drone flies inclined. For example, when the drone flies in perfectly horizontal position, the bank angle is zero. When it starts to turn its axis the bank angle starts to increase. When flying straight, planes bank angle is zero and when making left or right turns, they have a greater than zero bank angle.

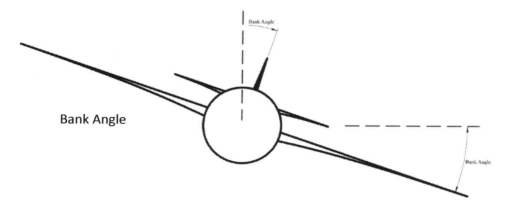

Bank Angle

Battery: Batteries provide essential power to the motors, receivers and controllers. For multirotors, the most commonly used batteries are Lithium Polymer (LiPo) types, as their energy efficiency is high. Usually 3-4 cell batteries are used, which provide currents of up to around 5000 mah (miliamperes - hour) capacity. To understand what mah means, consider this example: a 3000 mah battery will last 3 times longer than a 1000 mah battery. Think of the charge (or load) in Amperes and time as similar to velocity and time. Velocity x time = distance. Here the distance is mah, so in other words, it is the distance you can go for so many hours at a certain speed. As the speed (ampere or load you use) increases, the time will decrease because you have a certain defined mah limit (distance). The advantage of LiPo batteries are that it can discharge at a much faster rate than a normal battery. It is recommended to buy a few sets of batteries, so that, when the first set is discharged, you do not have to wait for flying your drone again, and while one charges in the recharger, you can use the other battery. Some intelligent batteries on newer models have sensors and they can calculate its distance from you versus amount of power to return. **Safety Note:** Lithium Batteries can catch fire and you must check the requirements of the battery manufacturer for safe usage.

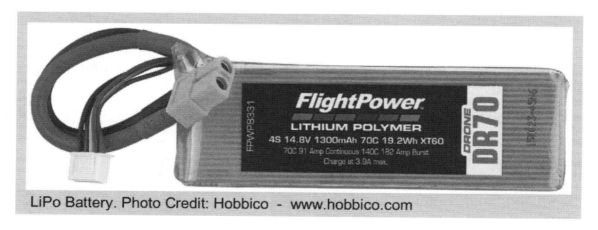

LiPo Battery. Photo Credit: Hobbico - www.hobbico.com

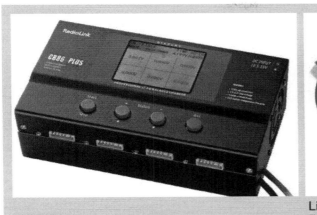

Battery Charger. Photo Credit: Radiolink
www.radiolink.com.cn

LiPo Battery. The values of mAh, Cell, Volt,
Discharge and Burst change from one battery to
another depending on the need.
Photo Credit: GetFPV - www.getfpv.com

BEC: Acronym for Battery Eliminator Circuit, also known as voltage regulator, and it is a circuit designed to eliminate the need for multiple batteries, by distributing power to other circuitry. Historically this was used for distributing power to battery driven equipment from mains electricity. For drones, it is part of the ESC or radio control receiver or power distribution board. BEC can also adjust the voltage amount, to various parts of the drone. If the drone has more than usual servos and electric circuitry, which requires more current than a single BEC can supply, a second BEC can be used.

Camera: Obviously the most frequent use of drones is to record aerial videos or take aerial photographs for many different reasons, therefore cameras are very important components of drones. Cameras may come with the drone when purchased, as built in cameras, or they may be external cameras that can be attached later. External cameras give more flexibility and control about the camera selection and they have their own batteries. The compatibility of camera and drone and your photo and video needs determine which type of camera or drone to choose. A camera with a remote shutter is also desirable. Some drone and camera systems come with phone apps that you can download to your phone and manage directly the video recording from your phone.

Photo Credit: UAV Factory
www.uavfactory.com

Photo Credit: Peau Productions
www.peauproductions.com

Photo Credit: UAV Solutions
www.uavsolutions.com

Catapult Launching: This is one of the methods to launch airplane drones, because airplanes need initial speed in order to fly. Catapults are used in order to throw airplanes into the air easily and quickly, where there might not be enough distance to speed up, or the drone might not have the gear to speed up (which saves weight and control systems). Catapult launched airplanes will need additional reinforcement in order to withstand the throwing force from catapult.

Photo Credits: UAV Factory - www.uavfactory.com

Dead Reckoning: To estimate the position only by using data from internal sensors, rather than GPS, relying on a previously known position. This is useful when there is interference with receiving data from GPS, such as indoors or near high structures or tunnels. Based on the last known position of the aircraft, (this could also be a land or sea vehicle), its position can be calculated only by knowing the distance and direction travelled since that last known point, which is measured by Inertial Navigation System, which include accelerometer, gyroscope. Because of relying on only one previously known point, dead reckoning calculations are subject to cumulative errors.

DGPS (Differential Global Positioning System): This is an enhancement to GPS, which provides improved accuracy, down to several centimeters, in comparison to meters for regular GPS. DGPS uses a local base station for reference, to enhance accuracy. The signals from base station is received by a GPS receiver on the drone. DGPS provides differential corrections to a GPS receiver. This improves accuracy and monitors integrity of information sent by GPS satellites.

Drift: Drone moving in a direction other than intended. Drift may be caused by winds or sensors that are not adjusted well.

Electro Optical Sensor: These are sensors that convert the quantity of light into electrical signals. These are always part of larger systems and used in a large variety of applications, wherever the light must be converted to signals, from position sensors, to smart phone screens.

Elevator: See *Pitch*

ESC: Electronic Speed Controllers, manage the speed and direction of motors. Each motor has one ESC that supply power and current about how fast it should turn. This ultimately means ESCs control in what direction, and how fast the drone moves, how it turns or accelerates and its direction. ESC may include BEC (Battery Eliminator Circuit) which distributes power to more than one circuit and LVC (Low Voltage Cutoff), in cases where the Voltage drops below a certain level. ESCs draw their power from the power distribution board and then transfer this power to motors. Therefore ESC must be able to provide enough current to the motors that the motors can handle. By varying the timing and the amount of current, the ESCs control the speed and direction of the motors. Most ESCs that are used for hobby grade drones are usually up to 30 amps. Also the refresh rate, which means, how many times per second the ESC checks for instructions from the flight controller, in other words, the speed of the ESC, can make difference in how well you can control your drone. ESC converts the DC current of the battery into three phase AC signals for the Brushless DC Motors. These continuous signals produce continuous changes in magnetic field of the motor and create rotation. The flight controller sends a signal to the ESC telling it how fast to turn motor. There is a circuit in the ESC that converts the signal from the controller into the much more powerful three phase signals need by the motor. In the figure below, the flight controller and power distribution board are not shown for clarity.

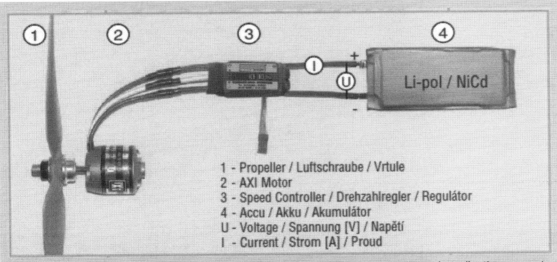

1 - Propeller / Luftschraube / Vrtule
2 - AXI Motor
3 - Speed Controller / Drehzahlregler / Regulátor
4 - Accu / Akku / Akumulátor
U - Voltage / Spannung [V] / Napětí
I - Current / Strom [A] / Proud

Electronic Speed Controller (shown as #3), controls how fast each motor turns, by adjusting current that goes to motor. Each motor turning at different controlled speeds, then controls the movements of the drone. Here the ESC takes its directives from the flight controller (also known as autopilot, not shown in this photo, for clarity) Photo Credit: AXi Model Motors— www.modelmotors.cz

A 4 in 1 ESC, which means, 4 pieces of ESC circuit are integrated to one board, which makes installation tidier and users do not need to solder multiple ESCs, and save effort. This also saves weight as it is lighter than using 4 separate ESCs. This is used on a race Quadcopter. Source: Maytech Photo Credit: Maytech — www.maytech.cn

ESC. Photo Credit: Kontronik Sobek Drives www.sobek-drives.de

ESC. Photo Credit: Maytech www.maytech.cn

Flight Controller: Also called the autopilot, or control board, this is the brain of the drone. It is an integrated circuit, which includes sensors, microprocessors, and input output pins. Flight controller not only directs drone where and how to fly, but it also provides stabilizing, hovering ability in the air, by counterbalancing the effects of wind. Levels of stabilization of drone can also be adjusted in some models by limiting the bank angle to high or low levels.

Advanced flight controllers can:

- Start hovering instantly in the air as soon as you release controls
- Autonomously take the drone from one point to another
- Return the drone to the base location in the event of a problem or command from the base
- Follow a moving person or object autonomously at a desired altitude, distance, bearing and angle

- Orbit an object
- Target a certain location and altitude
- Avoid obstacles in the air autonomously, indoors or outdoors, by also using cameras and ultrasonic sensors
- Focus camera gimbal at a point of interest, while the operator directs the drone
- Perform waypoint navigation, where the drone can travel along specified waypoints for a desired travel path
- Take panoramic photos of a target by directing the flight controls and camera gimbal accordingly

Although we focus our attention on drones here, it must be noted that autopilots can also guide manned aircraft. These can release pilots of tedious high altitude cruising tasks. Some can even do very precise maneuvers, such as landing an aircraft in zero visibility weather.

The chart below summarizes the direction of flow of commands in a drone. It can easily be seen that the flight controller lies in the center of everything:

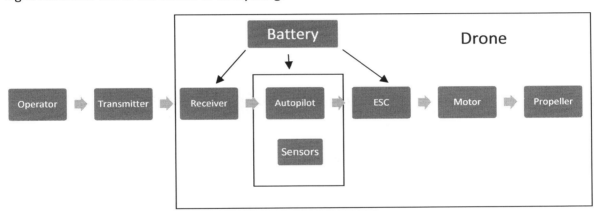

Flight controller receives information from two sources: First, it receives commands from the transmitter on the ground that is used by the drone operator, and second, it constantly receives feedback from gyro and accelerometer (which is in INS), plus other sensors if any, that it contains. With the help of INS, the autopilot constantly manages the hovering, tilting and speed of the drone as required. Other sensors may include barometer, which helps autopilot to keep altitude still or be adjusted as required, air speed sensor, ultrasonic sensor, which serves like a radar by reflecting sound waves on surrounding surfaces especially indoors to determine position, magnetometer, which measures magnetic field, GPS sensors, which helps to position the drone whenever satellite data is not obstructed.

After the combination of these two main sources of information are processed, organized into useful signals by using attitude estimation algorithms to communicate to multiple ESCs, the resulting commands are sent to each ESC at the right moment in a very fast ongoing process, in order to adjust the turn of the motors via signals. The signal from the flight controller is usually a PWM (pulse-width-modulated) waveform although other types of signal can also be used with some ESCs.

For hobbyist drones, there are different types of flight controllers, for different styles of flight. The three different styles of flight are Sports flying, Aerial Video Recording and Autonomous flying. Each of these require different set of functions from an autopilot. For example, if the purpose is to record videos, the flight controller must be capable of providing a very smooth flight. For autonomous flight style, the autopilot must be capable of performing relatively complex autonomous tasks. For the sports or racing mode, he flight controller must be able to handle high speed maneuvers better. Still it must not be forgotten that, flight controllers are configurable, and one flight controller may be set up to different configurations for different flight missions.

As a note for D-I-Y hobbyists, building a system to do all these is very difficult even for a person that has years of experience in programming, flight and electronics. If you are interested in developing or participating in building autopilots, we suggest that you participate in one of the open source autopilot projects.

Examples of such projects and code repositories are:

ardupilot.org   autoquad.org   dronecode.org
openpilot.org   pixhawk.org   px4.io
smaccmpilot.org   github.com/multiwii

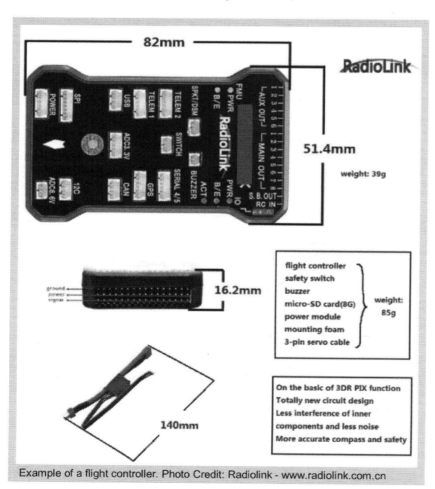

Example of a flight controller. Photo Credit: Radiolink - www.radiolink.com.cn

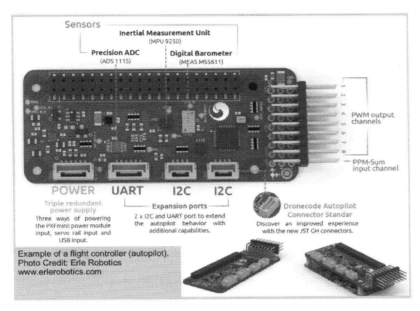

Example of a flight controller (autopilot).
Photo Credit: Erle Robotics
www.erlerobotics.com

Example of a flight controller (autopilot). Photo Credit: Erle Robotics - www.erlerobotics.com

Flight Logs: It is the form that is filled by the pilot, to keep a record of the flight information. Some advanced drone models can log and remember the flight details and do this automatically, by recording flight data from internal sensors such as flight distance, time, location, route can be logged in using GPS sensors. These and many other information about flight can be entered manually to keep logs. Some companies offer automated fleet management programs to track data.

Flight Time: It is the total time that a drone can spend in the air. Most available and popular drones for general consumer market have flight time range anywhere between 10-25 minutes. The drone operator must therefore carefully estimate the furthest distance that the drone can travel before losing contact and having enough time to return to the start point. Some drones can automatically estimate this based on their GPS sensor information. When doing flight time calculations, the ascend and descend rate of drones must also be taken into account. Note that for a tethered drone, the flight time could be hours or days, as the power can be transmitted through cable.

FPV Flying: Stands for First Person View Flying, it means to have an on board camera that enables to see things that are captured by the drone camera in real time.

FPV Monitor. Photo Credit: GetFPV - www.getfpv.com

Frame: This is the actual body of a drone, the skeleton that holds all the electronics parts, motors and propellers together. It can be of Plastic, PCB, Aluminum, Carbon Fiber, Foam or even wood. It must be of strong and light material. Carbon fiber is a tough and lightweight material and therefore suitable however, it impedes the radio signals. Wood has the advantage of being cheap and it can easily be replaced. Aluminum is also easily replaced, inexpensive and easily workable. Some frames are foldable, for ease of transporting.

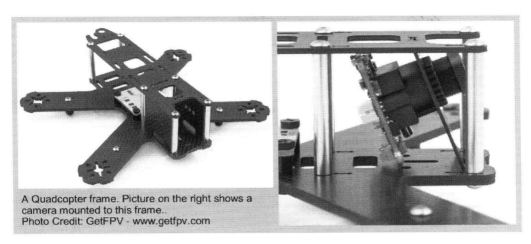

A Quadcopter frame. Picture on the right shows a
camera mounted to this frame..
Photo Credit: GetFPV - www.getfpv.com

A foldable frame. Photo Credit: Rosewhite - www.rosewhite.de

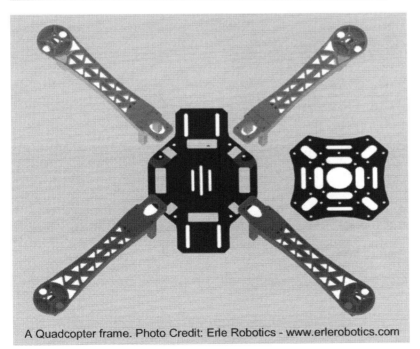

A Quadcopter frame. Photo Credit: Erle Robotics - www.erlerobotics.com

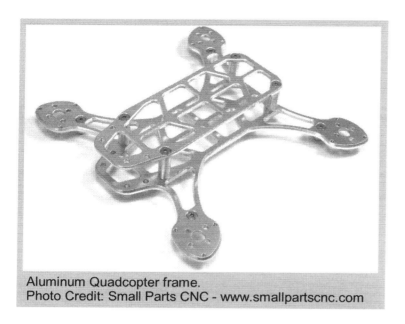
Aluminum Quadcopter frame.
Photo Credit: Small Parts CNC - www.smallpartscnc.com

Fuselage: The main body of an aircraft where wings and everything else is attached and placed into.

GCS: Stands for Ground Control Station, GCS are used to control advanced grade drones, such as military or mapping drones, by the operator from the ground. In addition to controls in a transmitter, these also have a laptop and special software to perform different operations for control and information processing such as maps, target detection and more functionalities with respect to a transmitter. Unlike transmitters, ground control stations may have unlimited range. GCS can control more than one UAV.

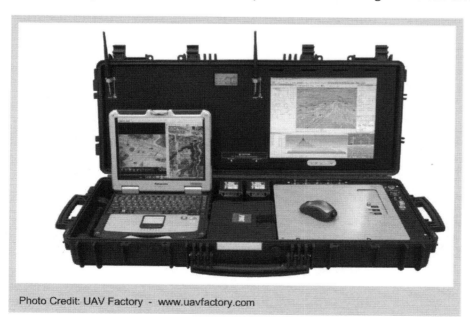
Photo Credit: UAV Factory - www.uavfactory.com

Geofence: A geofence is a virtual barrier that the drone must observe, for a real geographic area. It is imposed by the program as a radius from a certain start point, or, could be defined manually preflight,

14

such as setting airports or school areas as no fly zones, in order to keep the drone within those boundaries defined by the geofence, and prevent it from entering into no fly zones.

Geospatial Data: Data that has geographical component. For example, the data that is collected by drones from the surface of earth for mapping operations is geospatial data.

Gimbals: These hold the cameras in place and attached to frame of the drone and used to keep the camera stable in order to take smooth videos and pictures. They are also used to control the direction of cameras. Gimbals are 3 axis which means it can direct camera or counterbalance the negative movement effects of wind or shaky movements you cause in all axes. Even if the drone shakes considerably, gimbals can ensure that the camera stays at level surface, as the sensors inside can correct the orientation many times per second. Therefore gimbals also have motors to control and counter all these movements. If you only want to take photos and not videos, you may not need a Gimbal, and eliminate the extra weight and cost. In different flight modes, the gimbal is controlled by autopilot.

Photo Credit: Erle Robotics
www.erlerobotics.com

Photo Credit: Rosewhite - www.rosewhite.de

Global Positioning System: See GPS.

GPS: Acronym of Global Positioning System. It is a navigation system, where, location and time information about anywhere on the planet can be seen, as soon as the receiving device sees at least four satellites. On drones, the GPS chip communicates the location of the drone to controller and serves as one of the sensors of the drone. This chip also records the drone's starting position, and can enable flight controller to return the drone to that start position autonomously. This chip is also necessary to automatically hold the drone in stable position horizontally, such as adjusting its position automatically to compensate movements caused by wind, when the operator wants to hold it in stable location.

Ground Control Station: See GCS.

Gyroscope: Also called "gyros" in short, these are the sensors that measure rate of rotation of the drone about an axis, in degrees per second. Therefore the term you often see on multirotor packages as

"3 axis gyroscope" means, that gyroscope is able to measure rotation in all possible three axis, pitch, yaw and roll. Gyroscope helps in keeping the balance of the drone along with the accelerometer. Both accelerometer and gyroscope can be combined as one inertial sensor or IMU (Inertial Measurement Unit), which is a component of Inertial Measurement System (INS). For photo and charts, see INS.

IMU: Inertial Measurement Units are measuring devices that act like a combination of accelerometer and a gyroscope, in order to give complete information about drone's position, orientation, speed, acceleration. IMUs can also contain magnetometer, in order to also give information about magnetic field. By using IMUs, the drone can also track its position only depending on sensor input from accelerometer and gyroscope, without the need of GPS. This is an especially useful concept where GPS signals are not available, which is known as dead reckoning. As dead reckoning is subject to cumulative errors, the sensitivity of measurements provided by IMU is important.

INS: Acronym for Inertial Navigation System. It is a system that calculates position without the need for external position aid, such as GPS. It contains IMU.

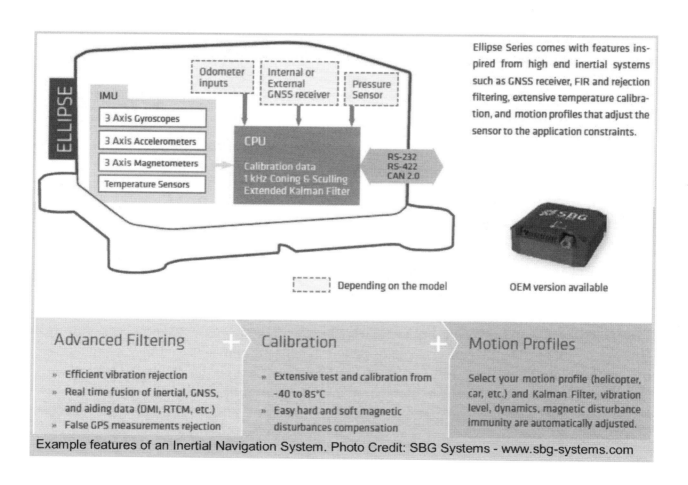

Example features of an Inertial Navigation System. Photo Credit: SBG Systems - www.sbg-systems.com

ACCURACY (RMS)

360° sensing in all axes, no mounting limitation

Model	A	E/N	D
Roll / Pitch	0.2°	0.2°	0.1° / 0.05° (PPK)
Heading	0.8° Magnetometers*	< 0.5° GPS**	< 0.2° Dual GPS*** (> 1 m baseline)
Velocity***	-	0.1 m/s	0.03 m/s
Position***	-	2 m	Single point L1/L2: 1.2 m
			SBAS: 0.6 m
			DGPS: 0.4 m
			RTK: 2 cm + 2 ppm (option)
			PPK: 1 cm (option)

Heave accuracy	10 cm or 10%	
Heave period	Up to 15 s	Automatically adjusts to the wave period

*Under homogenous magnetic field
** Under regular acceleration, or automotive motion
*** Under good GNSS availability
PPK = Post-processing Kinematic. Post-processing with Inertial Explorer®.

INTERFACES

Available data	Euler angles, quaternion, velocity, position, heave, calibrated sensor data, delta angles & velocity, barometric data, status, GPS data, UTC time, GPS raw data (Post-processing), etc.
Aiding sensors	GNSS, Odometer (DMI), RTCM
Output rate	Up to 200 Hz
Main Serial Interface	RS-232, RS-422, USB - up to 921,600 bps
Serial protocols	Binary eCom protocol, NMEA, ASCII, TSS
CAN interface	CAN 2.0A/B - up to 1 Mbit/s
Pulses	Inputs: Events, PPS, DMI (Direction or quadrature)
	Outputs: Synchronization (PPS), Virtual DMI
	Model A & N: 2 inputs / 1 output
	Model E: 4 inputs / 2 outputs
	Model D: 3 inputs / 2 outputs

INTERNAL GNSS

Engine, update rate	Model N: 72-channel, 10 Hz, L1 C/A GPS, GLONASS, QZSS, BeiDou, SBAS
	Model D: 120-channel, 5 Hz STD: GPS L1/L2/L2C, SBAS, QZSS Option: GLONASS, Galileo, Beidou
Cold start / Hot start	Model N: 26 s / < 1 s
	Model D: < 50 s / < 35 s

MECHANICAL

		Box	OEM model
Size	models A/E/N:	46 x 45 x 24 mm 1.8 x 1.77 x 0.9 "	34 x 34 x 13 mm 1.34 x 1.34 x 0.51 "
	model D:	87 x 67 x 31.5 mm 3.43 x 2.64 x 1.24 "	-
Weight		A: 45 g / 0.1 lb	12 g / 0.02 lb
		N: 47 g / 0.1 lb	12 g / 0.02 lb
		E: 49 g / 0.1 lb	12 g / 0.02 lb
		D: 180 g / 0.4 lb	-
IP Rating		IP68	-

All parameters apply to full specified temperature range, unless otherwise stated. Full specifications can be found in the Ellipse User Manual available upon request.

PRODUCT CODE
* standard product options

ELLIPSE-#-G#A#-##

MODEL
A: AHRS
E: Externally Aided INS
N: INS with integrated GNSS
D: INS with integrated dual antenna GNSS

GYROSCOPE
2: 100 °/s
3: 200 °/s
4: 450 °/s *
5: 1,000 °/s

ACCELEROMETER
2: 8 g *
3: 16 g

PACKAGING
B1 *Box* *
RS-232/422
B2 *Box*
RS-232 + CAN
L1 *OEM*
TTL
L2 *OEM*
RS-232/422 + CAN

SENSORS

	Accelerometers	Gyroscopes	Magnetometers
Range	± 8 g	± 450 °/s	± 8 Gauss
Gain stability	< 0.1 %	< 0.05 %	< 0.5 %
Non-linearity	< 0.2 % FS	< 0.05 % FS	< 0.1 % FS
Bias stability	± 5 mg	± 0.2 °/s	± 0.5 mGauss
Random walk/ Noise density	100 µg/√Hz (X,Y) 150 µg/√Hz (Z)	0.18 °/√hr	200 µg/√Hz
Bias in-run instability*	20 µg	8 °/h	-
VRE	7 mg/g^2 RMS	0.001 °/s/g^2 RMS	-
Alignment error	< 0.05 °	< 0.05 °	< 0.1 °
Bandwidth	250 Hz	133 Hz	110 Hz

* Allan Variance, @ 25 °C

PRESSURE SENSOR (models N & E)

Resolution	1.2 Pa / 10 cm / 0.3 ft	
Pressure accuracy	± 50 Pa / ± 200 Pa	Relative / Absolute

ELECTRICAL & ENVIRONMENTAL

Input voltage	Model A/E/N: 5 - 36 V
	Model D: 9 - 36 V
Power consumption	Model A/E: < 460 mW
	Model N: < 650 mW
	Model D: < 2,500 mW
Specified temperature	Model A/E/N: -40 to 85 °C, -40 to 185 °F
	Model D: -40 to 75 °C, -40 to 167 °F
Shock limit	2,000 g
Operating vibration	3 g RMS (20 Hz to 2 kHz per MIL-STD 810G)
MTBF	50,000 hours

Example specifications of an Inertia Navigation System. Note the range and tolerances given for Accelerometer, Gyroscope and Magnetometer. With the help of its sensors, the INS not only guides the autopilot but also can help determine the position of the drone even when GPS data is not available, such as dead reckoning. Photo Credit: SBG Systems - www.sbg-systems.com

Example of an Inertial Navigation System. In addition to Accelerometer and Gyroscope, which is normally included in IMUs, these particular sensors also contain Magnetometer and Temperature Sensors. An IMU comes without Kalman filtering. As soon as there is an embedded filter (processing unit), it is called Attitude and Heading Reference System (AHRS), and if it embeds a GPS/GNSS receiver, it is called an Inertial Navigation System (INS). For visualization, one side of the smaller sensors is about a few centimeters here. Photo Credit: SBG Systems - www.sbg-systems.com

Initializing: This is done to initialize the gyroscope, before the flight, by placing the drone on a level surface and letting the gyro "understand" the level surface.

Kaman Filtering: This is an estimation algorithm that is used in inertial navigation sensors, which is used for producing more reliable results from the measurements of the sensor, in comparison to not using this algorithm.

KV Rating: This is used to rate the motors, which Motors are identified by their KV rating, which is a value that shows how fast a motor will rotate under a certain voltage. For a multirotor, a KV rating of up to 1000 can be suitable, whereas for more aggressive flight, such as action recording or acrobatic flights this value can be more. For example if a motor has a KV rating of 1000 rpm/V, then, under 7.4 V, the motor would rotate 1000x7.4 = 7400 rpm.

Landing: Multirotor or helicopter drones land and takeoff vertically, but planes land while decelarating and moving forward. For this reason, the landing gear of multirotors and helicopters are in the form of skids, while planes land on wheels.

The time for landing must be taken into account when calculating flight times. Landing takes place slower than ascending, due to the precise maneuvers needed.

Also, helicopter or multirotor drones must land slowly, or they will get caught in the downward pushing air that their own propellers generate. If the drone gets caught in this downward pushing air, which sucks the drone downward like vacuum, trying to throttle up to escape this will cause even a stronger vacuum, pulling the drone down even more.

Landing Gear: For multirotors, these are bent shaped skids that protect drone in case of rough landings. They come in different shapes and sizes depending on the need. The main idea is to absorb the shock of rough landing so as to minimize the damage to the drone. Some landing gears are fixed and others are retractable in order to allow the camera to turn and also not to block the view when recording videos or taking photos. They may be of carbon, aluminum or special plastic material.

Retractable gears come with servos for retraction. While fixed landing gears are very cheap, retractable ones might be considerably more expensive.

Landing Gear. Photo Credits: Rosewhite - www.rosewhite.de

Led Lights: Led lights are sometimes used on drones for illumination of the drone or its surroundings, as well as aesthetic reasons during night time flights. It is also required by many drone race organizers.

LED Lights (top)

LED Lights attached to frame. See also the PDB for the lights in the middle.

Photo Credit: GetFPV www.getfpv.com

LIDAR: Light Detection and Ranging is a remote sensing technology, which uses many pairs of laser beam reflections to measure distance, that are used for measurement and positioning and making maps, surveys and 3d images of buildings, natural resources, infrastructures or other features on the ground. With use of LIDAR, high resolution maps in geomatics, contour mapping, geography, archeology, forestry and any similar field is possible.

Lift: The resultant upwards force created by the propellers in order to lift the aircraft. This is achieved either through creating pressure differential at the faces of the wings in case of a plane, or pushing the air down in case of a helicopter or multirotor. Number of propellers, size of motors, battery power all affect the lift force.

LVC: Stands for Low Voltage Cutoff, and part of ESC. This functions when the charge of batteries drop and the voltage is cut to drive motors and supplied to steering servos, in order to safely return the drone to the base.

Motors: Motors are powered by the batteries and make the propellers turn. Motors in a multirotor drone do not draw power directly through the battery. The power is transmitted through ESCs, for each motor, and the ESCs get their power from the power distribution board, which is powered by the battery. Larger drones need larger motors and larger batteries. Motors are identified by their KV rating and thrust.

Photo Credits: AXi Model Motors - www.modelmotors.cz

There are different types of motors:

Brushed Motor: This can be considered as the basic type of motor, and the control is as easy as turning a current switch on and off, which enables a certain motor speed. In this type of motor, the permanent magnets are stationary, and the electromagnet is moving. The current is applied to brushes, from the battery, and these brushes, which are stationary, physically touch to commutator, which is moving. So this physical contact of a stationary piece to a moving, constantly turning piece. creates friction and heat and causes the brushes to wear out by time. When the commutator takes the current from the brush, it sends this current to the coil of wires which creates a magnetic field, and the permanent magnets make these coils rotate by pushing and pulling them according to their position. The commutator can be considered as a switch, when it turns, the polarity of magnetic field is switched each time a coil passes a magnet. It is this constant switching of magnetic field that creates the turning force. The disadvantages of brushed motors are that they are not very efficient due to friction and heat loss, and brushes and commutators wear out in time. This means, they must be maintained more often. The advantage is that they can be constructed easier than brushless motors and the initial cost of building this motor is lower. Brushed DC motors have two connectors, for positive and negative. To change the direction of rotation, the only thing needs to be done is changing positive and negative from the switch or battery connection.

Brushless Motor: In a brushless motor, as opposed to a brushed motor, electromagnets do not move, but the permanent magnets move, by either being on the inside (inrunner motor) or outside (outrunner motor) of the electromagnets. The movement can be achieved when the magnetic field generated by the electromagnets pushes and pulls the permanent magnets. Therefore at no point for this push and pull process there needs to be friction between pieces which increases the efficiency of these motors, in comparison to a brushed motor. This also means that considerably less heat is generated by these motors. The sequence of providing current and sensing where the permanent magnets are done through ESC and sensors and it is more complicated than constructing a brushed motor. Therefore, the initial effort and cost of building a brushless motor is higher but later it needs less maintenance as there is no wearing brushes due to friction. Most drones today use brushless motors, but very small ones, such as the ones that are less than 6" diameter, may use brushed motors due to simpler controls and lower cost. Brushless DC motors have three connectors which are controlled by ESC. The ESC converts the DC current of battery into three phase AC signals for the motors, which continuously change magnetic field and produce rotation. Therefore, if a brushless motor is connected directly to an unchanging DC power source, the motor will not be able to change magnetic fields and instead of turning it will short circuit and melt.

Photo Credit: Kontronik Sobek Drives
www.sobek-drives.de

Inrunner Brushless Motor: The turning piece is the inner shaft and therefore it can turn very fast but produces little torque and therefore needs a gearbox in order to increase the torque delivered. The inner turning shaft is the permanent magnet as this is a brushless motor and the fixed coils are at the outer casing.

Outrunner Brushless motor: This type of motor spins its outer shell, so although slower, it is able to produce much more torque than an inrunner motor. Directly producing a lot of torque in the beginning eliminates the use of gearbox, which reduces weight, noise, complexity, efficiency loss and makes this type of motor suitable for running all types of drone propellers. The outer turning shaft is the permanent magnet as this is a brushless motor.

A typical data sheet for a brushless multirotor motor by manufacturer may look like below:

Throttle (%)	Volts (V)	Propeller (inch)	RPM	Watts (W)	Thrust (g)	Current (A)	Efficiency (g/W)	Operating Temp. (g)
50	22.2	8	1100	40	700	1.7	17.5	39
75	22.2	9	1600	91	1450	3.9	15.9	40
100	22.2	9	1950	141	2050	6.0	14.5	41

One thing to note about this data is that the efficiency (grams / Watts), goes down, as you increase the throttle to 100%.

Prop	Volt	Watt	Amp	Thrust (g)	Throttle %	Efficiency (g/W)
HQ 4x4.5 Bullnose	16.5	90.8	5.5	265	50	2.92
Nylon-Glass Fiber	15.8	331.8	21	669	100	2.02
GemFan 5x3	16.5	57.8	3.5	254	50	4.40
Carbon Nylon	16.1	206.1	12.8	660	100	3.20
HQ 5x4.5	16.8	86.7	5.16	342	50	3.95
Nylon-Glass Fiber	16.5	316.1	19.16	802	100	2.54
GemFan 5x4.5 Bullnose	16.6	114.5	6.9	376	50	3.28
Nylon-Glass Fiber	16.4	373.9	22.8	866	100	2.32
HQ 5x4x3	16.6	112.9	6.8	426	50	3.77
Nylon	16.4	406.7	24.8	962	100	2.37

Brushless Motor. The chart is a typical data sheet that shows different efficiency values for various propeller sizes, voltages and throttle percentages. Photo Credits: GetFPV - www.getfpv.com

Mounting Plate: This is the center piece of a multirotor, that holds the flight controller, battery, receiver, camera. Around the mounting plate, the arms of the multirotor are attached. The term mounting plate is also used for the pieces that attach motors to the end of the arms of the drone.

Operational Range: How far the drone can operate safely, measured in horizontal, without losing contact with the transmitter, and allow the drone to return starting point without running out of batteries or fuel.

Parachute: Using a parachute adds to the payload but it is a good idea in order to prevent crashes in case of a malfunction or other unforeseen circumstance. The parachute slows down the fall and prevents damaging crashes which may damage the drone itself, property or people.

Photo Credit: Mars Parachutes
www.MARSparachutes.com

Photo Credit: Mars Parachutes
www.MARSparachutes.com

Payload: It is a term to express how much weight the drone carries, in addition to its own weight. The larger the propellers of a multirotor drone, the more payload capacity the drone has. But it must not be forgotten that larger propeller can only be turned efficiently with larger and heavier motors and batteries. The total weight of a drone is one of the biggest factors effecting its performance, as far as the flight time or maneuverability. For example a reduction of %20 of total weight of a drone's weight, may mean a %20 increase in flight time, given the same battery capacity, propeller and motor. Therefore, the payload must be chosen carefully.

Phone Apps: Many drone manufacturers have come up with apps that make control of drones via smart phones. Through these apps,

- the drone can be controlled just by tilting the phone or touch screen commands
- setting of many functions for autopilot for flight modes can be made
- the camera and gimbal can be controlled
- real time, FPV viewing of camera o screen and sharing photos or videos instantly is possible
- real time flight parameters can be seen on screen

- detailed records of flight can be kept
- simulated flights can be flown
- many other functions not listed here, depending on the manufacturer

PID Settings: PID stands for proportional, integral and derivative. It is used in industry to control and correct processes based on the error values obtained, acting like a constant feedback loop (or control loop). For drones, this principal is used for correcting how the drone flies when it tilts by an external force such as wind, or reacts too much to the operator's move commands or the center of gravity is off. PID settings should be tuned before flight. It is done as a trial and error process when you tune them for your drone. For example if in the beginning the P setting is too low, the multirotor will roll, yaw and pitch a lot and will be slow to react. It should be increased until it becomes stable. However turning it up after a critical point will make it oscillate with high frequency, so if you want a smooth flight, refrain from increasing this value too much. P setting is the main value. I setting is to overcome external effects like correcting for wind. If I value is too high, the drone will be hard to control and rotate in all axis. And finally D setting is for controlling the response of the drone when you make aggressive moves such as action recording, so for these kind of flights, D setting should be high. In other words, this will increase how fast the drone reacts to the user. P is for present errors, I is for the total of past errors and D is for the estimated future errors. PID controls work by measuring the actual angle feedback from the gyro sensors and comparing it to the desired angle, for all three axes of turning. So this is the core software algorithm of the flight controller.

Pitch: (Also called Elevator) Tilting the multirotor forward or backward. This is achieved by turning blades faster on one side and slowing down at the other. (For the pitch of propeller, see propeller). The figure below shows the turning axis for Pitch, Roll and Yaw.

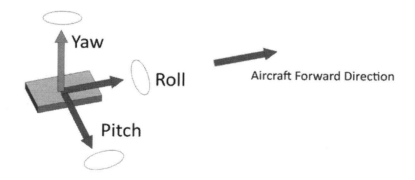

Power Distribution Board: These flat boards contain soldering points for cables in order to distribute power within the drone between different electronics. PDBs have input and output points, or positive and negative terminals which are neatly connected to each other. Some drone frames may already have power distribution board build into them except carbon fiber frames, because carbon fiber conducts electricity and the board must be mounted separately in this case in order to avoid short circuits. Simply by connecting the cables, without a PDB, power can also be distributed, but the board helps keep everything tidy. PDBs come in different sizes and current ratings, and different number of connection points, and some include BECs, (voltage regulators). For very high currents PDBs are not suitable but for a normal hobby grade drone, it is unlikely that there will be currents that PDBs cannot handle.

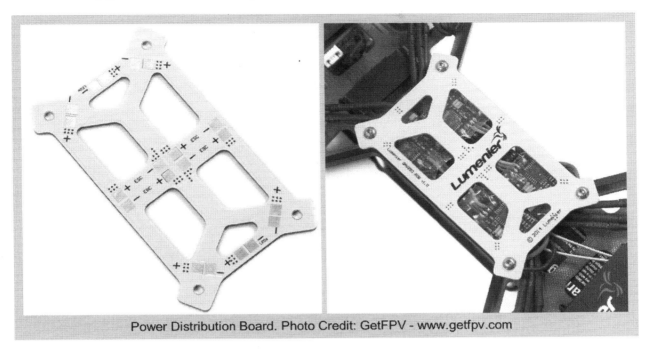

Power Distribution Board. Photo Credit: GetFPV - www.getfpv.com

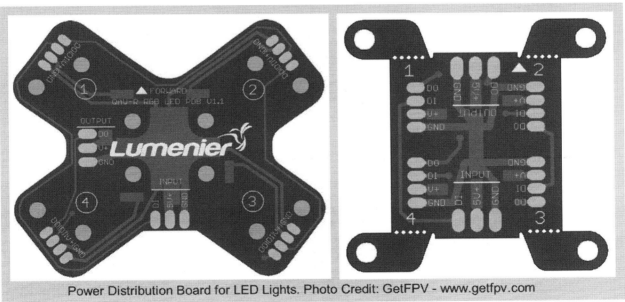

Power Distribution Board for LED Lights. Photo Credit: GetFPV - www.getfpv.com

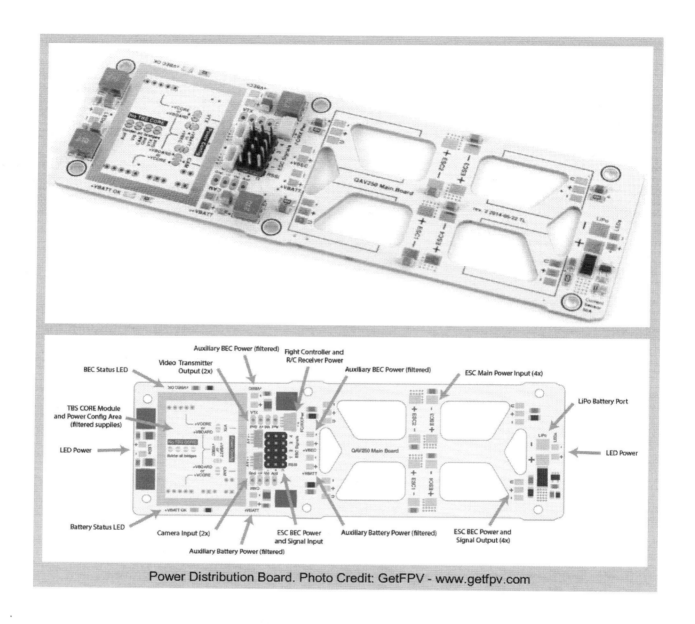

Power Distribution Board. Photo Credit: GetFPV - www.getfpv.com

Propellers: Propellers carry the drone by turning and pushing air downwards. They are delivered in Clockwise and Counterclockwise turning pairs. In other words, both pairs push the air downwards, but one pair turns CW and the other turns CCW. For a multirotor, buying CW and CCW propellers in different colors can help identifying them quickly on the drone in order to determine the orientation of the drone, as it is important to know the front side of the multirotor during flight and from ground. The frame colors will also help. Turning of propellers in CW and CCW direction enables rotating or tilting the drone, as explained by the graphics under the term pitch. In case of a plane, the propellers push the air backward to underside and upside faces of the wings so that the upper face has small pressure and lower face has high pressure which lifts the aircraft. A certain size propeller has a certain lifting capacity and beyond that, it cannot push the air efficiently. Propellers must be chosen as far as their size and pitch. The pitch of propeller is the angle which the propeller pushes or absorbs the air. The bigger the angle, the more air it can push but this means more power consumption. So for example a ship's propeller can adjust the push of the water just by changing angle of the blades, without changing the rotation speed. The twisted shape of the propellers is because of this reason, as the inner side of the propeller turns at a slower speed than the other edge of the propeller, and in order to get the same push or absorption throughout the blade, the inner side of the propeller has more pitch than the outer edge by transitioning in between. Propellers must be attached in self tightening direction, so that it will not turn loose in time.

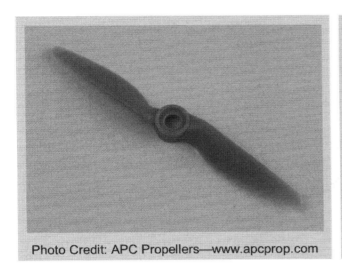

Photo Credit: APC Propellers—www.apcprop.com

Photo Credit: Maytech — www.maytech.cn

Propeller Adapters: These are attached on to the motors, in order to connect the propeller to the motor. Also called adapter rings.

Photo Credit: APC Propellers—www.apcprop.com

Propeller Balance: Propellers are balanced to make sure that the opposite blades have equal weight / moment with respect to center. The heavier side will hang lower, so it can be understood that the propeller is not balanced. Before placing the propeller on the balancer, it must be made sure that the shaft is perfectly horizontal. After balancing, if one side of the propeller is found to be lighter, a thin tape can be added to that side, to make it heavier, until the imbalance is taken care of. An alternative would be to drill small holes at the lighter side and fill with heavy material, or paint the lighter side with compatible paint with enough number of coatings to make that side heavier. If the imbalance is very little, the heavier side can be sanded too. The balancing must continue until the blade can stay stable at any position, not necessarily just horizontal, without any side hanging lower, or tilting smaller than a satisfactory tolerance. Note that the hub of the propeller can be out of balance too. In that case, some glue might be added to the lighter side of the hub.

The propellers coming out of the factory may not always be balanced and it is good practice to balance them before using and even after starting to use, from time to time. Imbalanced propellers may vibrate and harm the mechanical and electronic system of the drone, bearings inside motors over time. An imbalanced propeller will also draw more current because of its vibration, therefore reducing the flight time. The imbalance of propellers will be noticed more at higher turning speeds.

Photo Credit: Dubro — www.dubro.com

Propeller Guards: These are attached to the frame, in order to surround the propellers and protect propellers to contact other objects. In many cases, propeller guards are built as part of the frame and not separately attached. In some models, if propeller guards are attached, obstacle sensing system cannot work, as the guards may interfere with the sensors.

Real Time Kinematic (RTK) Technique: This is a system that is used for positioning the drone with the help of a base station. The base station has a precisely known location and it can be used to communicate with multiple devices at a distance of up to 10-20 kilometers.

Receiver: See Transmitter and Receiver.

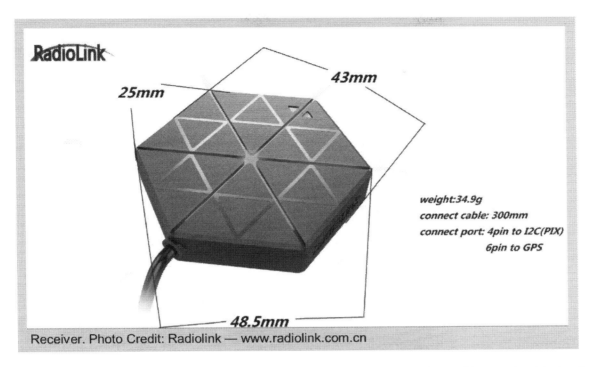

Receiver. Photo Credit: Radiolink — www.radiolink.com.cn

Roll: (Also called Aileron) Tilting the multirotor left or right. This is achieved by turning blades faster on one side and slowing down at the other. See graphic representation under pitch.

Rudder: See Yaw.

Sensors: Sensors provide useful information about the surroundings, or the drone's movement itself, in order to control the movements of the drone effectively. There are different types of sensors. In almost all drones, the most common sensors are, Accelerometers, Gyroscopes, which are parts of IMU. These are multiaxial, in order to enable drone to sense acceleration and rotation in all directions. There are other types of sensors, for other specific uses as well. For example, Ultrasonic Sensors are used to position the drone usually in indoor operations, by reflecting sound waves from surfaces which enables measuring distances and therefore positioning the robot. On the drone, these must be placed as far away as possible from noise or vibration sources such as propellers. Magnetometers measure the magnetic field strength and can be used to detect different minerals underground, or metallic underground piping. They are also used for detecting the compass position of the drone. Air pressure sensors (barometer) determine the height of the drone, based on the principal that the air pressure decreases as height increases. GPS system detects the position of the drone with respect to satellites and therefore give accurate position of the drone on earth. Lidar (Light Detection and Ranging) sensors, which use many pairs of laser beam reflections to measure distance, can be used for measurement and positioning and making maps and 3d images of buildings or other features on the ground. Heat sensors can have rare but specific uses in order to detect temperature differences such as locating people after a disaster.

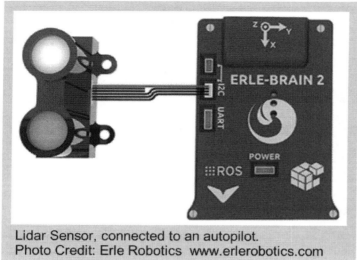

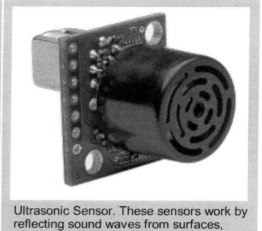

Lidar Sensor, connected to an autopilot.
Photo Credit: Erle Robotics www.erlerobotics.com

Ultrasonic Sensor. These sensors work by reflecting sound waves from surfaces, which enables measuring distances.
Photo Credit: Maxbotix www.maxbotix.com

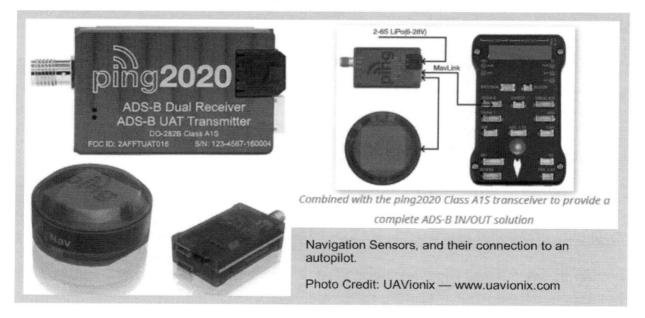

Combined with the ping2020 Class A1S transceiver to provide a complete ADS-B IN/OUT solution

Navigation Sensors, and their connection to an autopilot.

Photo Credit: UAVionix — www.uavionix.com

Shell: Shell is attached to frame as a cover to protect the central core of the drone from water or sun, at the same time improving appearance and aerodynamics. As other parts, it must be made of lightweight but durable material.

Speed: The speed of the drone in relation to the ground. Generally the speed for hobby grade drones can range anywhere from 15 to 50 miles per hour (roughly 25 to 75 km/h). Airplane drones can have much higher speeds.

Strafing: Move from left to right repeatedly without changing forward looking direction.

Telemetry: Telemetry refers to automatic collection of data and transferring of this to a remote location.

Tethered Drones: Tethered drones are tied to the ground which offers some advantages. These long lightweight cables can be used for power transmission so one of the main advantages is much longer flight time. Tying of the drone also ensures that the drone doesn't fly away and therefore allows lower skilled operators use their drones without the concern of losing the drone. Tethering can also offer immediate and fast data transmission to the ground from the drone. Having a drone in the air for long period of time, hours or even days, without worrying about the depletion of battery every 10 minutes, offer many advantages and new possibilities that cannot be done with non tethered drones. The main disadvantage of tethering is the risk of entanglement of cable to the drone. Only helicopters or multirotor drones can be tethered, it is not a suitable application for a plane due to high speeds and longer distances.

Throttle: Moving the multirotor up and down, without tilting or turning anything, just by increasing or decreasing the speed of propellers. In other words, it basically controls the level of power delivered to the motors. To visualize easier, throttle is similar to a gas pedal of a car. During lowering of the multirotor drone, doing things too fast may get the drone caught in the downward pushing air that was just created by the propellers and results in drone moving too fast downwards and even crash landings. Trying to throttle up in this situation will cause to vacuum be even greater.

Thrust: It is the indication of how much weight a motor can lift given a propeller size. Motors that are manufactured for multirotors can have specifications from manufacturers that list thrust values for different propeller sizes. When building a drone, the total weight of drone must be calculated first, then a payload weight is added, if any, finally a safety margin is added, and this will be the total resultant uplift force that is needed. And this value is divided by the number of propellers to find the load per propeller or load per motor. Based on this value, a motor can be chosen or drone dimensions, propeller and motor sizes can be adjusted, until satisfactory values are reached.

Tracking Antenna: Tracking Antenna autonomously tracks the moving UAV, by directing the antenna to the aircraft, using telemetry data. It is used together with ground control stations and for advanced drone systems and can have over 100 km range. Tracking antenna enables long range data transmission.

Photo Credit: UAV Factory - www.uavfactory.com

Transmitter and receiver: These are remote control devices used by the drone operators to control the drones from ground. The control of the hobby grade drones are typically achieved through transmitters and receivers using radio waves. Smart phone, tablet or game controller is also possible which use wifi or Bluetooth but they can control only shorter distances. Transmitters come with a receiver which can be first paired with the transmitter, and then, attached to either the motors directly, if manual control will be used, or the input channels of the autopilot, if the autopilot will be used. The typical radio wave length used to operate drones is 2.4 Gigahertz (GHz). Transmitter may come with drone purchase or may need to bought separately. Reaction times and interference free operation is very important when choosing transmitters and receivers, as well as making sure the drone and transmitter are compatible in the first place. On a typical radio transmitter, elevator trim, aileron trim, rudder trim, power indicator, power switch, landing gear switch can be seen. Important note: As the controls use radio waves, you must be aware that radio waves are subject to interference from other sources, and therefore this may cause temporary loss of drone control. So plan your flight accordingly and take your precautions.

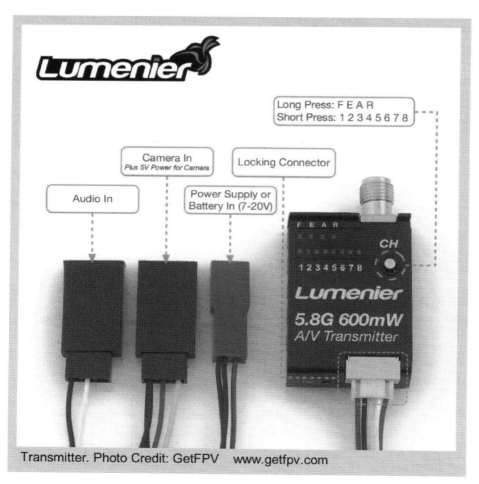

Transmitter. Photo Credit: GetFPV   www.getfpv.com

Transmitter. Photo Credit: Radiolink
www.radiolink.com.cn

Transmitter. Photo Credit: Radiolink
www.radiolink.com.cn

Trim: On a transmitter there are control rods that can be moved to control Throttle, Rudder, Aileron, Elevator moves, which are called as trim.

Visual Positioning: Using camera to locate the drone to help GPS or in the absence of GPS, such as indoors, and rely totally on it for navigation. Cameras can also make a map of drone's surroundings.

Visual Tracking: Tracking a moving object autonomously.

VTOL: Abbreviation for Vertical Takeoff and Landing. Helicopters and multirotors can perform this, while airplanes cannot (there are airplanes which can do this but it is exceptionally rare for airplanes).

Waypoint: These are coordinates, used to identify points. On the surface of earth, waypoints are two dimensional, such as longitude and latitude. In the air, it is three dimensional, with the addition of height dimension. With Waypoint Navigation, the drone can be navigated through predefined points for a desired path with autopilot. When a waypoint flight path is prepared, it must be remembered that the route needs to be clear of all obstructions, both horizontally and vertically, and the ascend and descend time of drone must also be taken into account for the distance versus time calculations.

Yaw: Also called rudder. Rotating the multirotor left or right in horizontal. This is achieved by spinning rotors turning in one direction faster and the other direction slower. See graphic representation under the term pitch.

Overview of building a hobby grade multirotor drone

This section, is especially directed to beginner to intermediate level hobbyists, to build a multirotor drone by themselves. In here, general information and main steps are given, but details are not provided.

If you are a beginner, it is strongly recommended that you review the previous section, where terms and concepts about drones are given, if you haven't done already. It is also our recommendation that before making your first drone, buy a small drone and learn how to fly drones. This will help save time and money for the future.

Important Note: In the US you must register your drone if it weighs over 0.55 lbs (250g). This should mean, even if you make your own drone, you must register it, if it is able to fly outdoors. Check with FAA and your state and local authorities. All regulations and safety guidelines still apply for a drone that you built yourself - actually even more, both during and after building.

Basic steps of building a multirotor drone and choosing parts:

- *Build or choose the frame:* For the frame, choose as light material as possible, while maintaining strength. You can also buy a ready built airframe or even 3D print. This is up to you but you must first consider the weight and how many and what size propellers can carry this. No matter what material you choose, proper cutting tools are needed if you build the frame yourself. The lifting capacity is dependent on the motors and the size of propellers. Frame must be able to fit those motors and propellers and the choosing of those are described below. So choosing the frame and motor and propellers go hand in hand as the size and weight of one affect the others. Apart from the size, frame must be strong enough. For example it must withstand the forces of rough landings. Also frames may flex under the force of motors that turns propellers in the opposite direction. Another thing to look is that the frame has as lower center of gravity as possible, to be more stable during turns. For transporting the drone in a small bag later, such as a backpack, or storing your drone, the foldability of the frame is also important. If you are building the frame, you can start by attach the arms to the mounting plate that will later hold the flight controller, receiver, power distribution board and the battery. It would be a stronger hold for the arms, if you use two mounting plates and sandwich the arms between the two plates, rather than just one center plate. The pieces to hold motors at the end of arms should also be attached. After determining the propeller size you can determine the size of your frame, by drawing the 4 propellers on paper or using a CAD software, giving them just enough space between each other so that they will not try to move the same air (note that this distance gets bigger as propellers get bigger), and giving your electronics and camera gimbal (or any payload if it will carry payload) enough space in the middle.

- *Choose Motors:* First of all it is highly recommended that you choose a motor specifically made for multirotor use. There are thousands of different brand and type combinations and in the beginning it may be confusing. Some motors come with their own propeller adapters and their own custom mounting pieces for easy installation to the frame.

The specifications when choosing motors include but may not be limited to the items below:

Kv Rating (rpm/v)
Compatibility of battery number of cells such as 2S, 3S, 4S etc...
Weight
Propeller Sizes compatibility
Maximum Surge value (Watts)
Mounting Hole Dimensions
Working Current, Maximum Current, No Load Current (In Amperes)
Shaft diameter
Output Wires
Thrust
Throttle
Voltage
Efficiency
Outside dimensions of Motor
Number of Stator Poles
Number of Poles (each)
Internal Resistance
Max Lipo Cell

For a typical specification sheet, see Motors in the previous section of the book.

When choosing motors, the main concern is the lifting capacity. The lifting capacity with different types and sizes of propellers are given. Calculate the total weight of all components plus secondary items like attaching camera with gimbal optionally. You can also attach a parachute, landing gear and a payload, which also add to the weight. As a rule of thumb, a quadcopter with 10 inch propellers, can carry up to 3-4 pounds of payload. This is not considering its own weight. Don't forget that we said quadcopter here. If your drone will have 6 propellers, or even 8, or more, then of course higher payload capacities can be achieved but things get more complicated. Especially if you are a beginner, we suggest that you start with nothing more than a quadcopter. Also add a safety margin on top of calculated total weight and choose the motors per that larger value by dividing the total weight with the number of motors. This safety margin is needed so that your drone can respond to your commands better when maneuvering or in case of emergency. You must do a little research on this and come up with a decision for number and dimension that you will feel comfortable with. Not only the lifting capacity of motors is to be considered but also current requirements too. For a drone, the most amount of current is drawn by motors and other components on the drone draw insignificant current in comparison. So the total of peak current of motors plus a safety margin, must be compatible between motors, ESCs, the battery and it must be something PDB can handle. Note that PDBs cannot handle high currents. In that case breakout cables should be used, which essentially means wiring all the ESC power lines with sufficiently thick wires.

- *Determine the size of propellers:* This very much goes together with selecting the motors to achieve the desired lifting capacity for a given size of propeller versus given motor capacity.

Propellers also must have enough space between each other depending on their size. Too close propellers will try to move the same air which is inefficient. Therefore propeller size affects the size of frame as discussed above.

- *Choose the flight controller:* Given your needs and what you need to do with your drone, you need to choose a flight controller. At a minimum, things to be considered when choosing are: the possibility to be configured for various multi rotor configurations, has enough sensors for your purpose, and possibly a straightforward setup process. The software should guide you through the processes such as selecting the multirotor type, motor layout and propeller direction, radio and ESC calibration.

Other features of flight controllers may include:

  - Assistant software for smart phones
  - Failsafe features such as auto return home or auto landing in case of emergencies
  - Take off assistance
  - Failure protection for motors
  - Supporting several inputs and receivers
  - Multiple control modes and intelligent switching between modes
  - Low voltage warnings
  - GPS module for accurate position stabilization, gimbal stabilization
  - Intelligent orientation control
  - Wireless PID tuning
  - Arm/disarm modes for safe motor starting
  - Support of PPM S-Bus or general receivers
  - On board USB for setup
  - Allowing of remote adjustment of gain

The specifications when choosing flight controllers can include but may not be limited to:

  - Weight
  - Dimensions
  - Microprocessor data
  - Sensors (gyroscope, accelerometer, magnetometer, barometer)
  - Supported multirotor types and configurations
  - Antenna Connector
  - Supported ESC output with refresh frequency (Hz)
  - Recommended transmitter
  - Working voltage range including input output voltages
  - Interfaces such as input and output data, ports
  - Recommended number of battery cells and burst current
  - Maximum and normal values of power consumption
  - GPS, Compass Ports
  - Operating temperature, hovering accuracy (meters)
  - Maximum yaw angular velocity (degrees /s)
  - Maximum tilt angle (degrees)

- Ascent and descent rate (meter/sec)
- Built in functions

For more, also see information in the first section of this book, under flight controller.

- *Choose and calibrate ESCs:* When choosing ESCs, you must choose all of them the same type, in order to get same response to controls. The main criteria you need to consider while choosing ESC is the current rating, in other words, it must be able to handle the current that the motor requires. Also as usual you must choose the lightest ESC as possible for your need. Note that for peak current calculations for your battery and PDB compatibleness, you must add the peak currents of motors, but not the current ratings of ESCs. The current rating of an ESC is a maximum rated value for that ESC and not the actual value of all motors will draw even at their peak instance. The current rating of ESC must always be higher than the peak current that a motor can draw as a safety margin so it is not the actual value and can be as high as it can, and therefore not relevant in peak current calculation.

The specifications when choosing ESCs can include but may not be limited to the items below:

- Current Rating (A)
- Voltage Range (V)
- BEC - if any
- Weight (including cables)
- Motor and Discharge Plugs and Wires
- Size
- Input Frequency
- Firmware

After all of the assembly of your drone is complete, the ESCs must be calibrated in order to make sure that they all respond to controls the same way, in other words they all know the high and low points of throttle range. Without knowing the throttle range, for example, you may start the throttle and as you go up one ESC may start and the other may not start yet, and this will of course make flight impossible, therefore this is an important step that must be done. The calibration is done by directly connecting the ESCs to the throttle through the receiver, and introducing the low and high points of the throttle to the ESC. For the further details of how to calibrate the ESCs, there are lot of videos on internet that you can watch, and we will skip the details.

- *Connect Motors to flight control system:* The flight controller must be setup in order to fully know the components of your drone. This is done via software and then it can be uploaded to your flight controller. The motors are connected to the flight control system via ESC. When connecting ESC to the motor, any one of the three wires coming out of ESC can be connected to the motor. On the other side of ESC, there are wires to connect to power distribution board and to the flight controller, where the ESC draws the power from the battery, and with the signals that are sent from the flight controller, adjusts the amount of current that goes to the motor. The black wires from ESC should be connected to the negative pins and red wires to positive pins of power distribution board, just like the black and red wires from the battery. Power cables between motor and controllers must be as short as possible. Also see the photos under the ESC description in the first section of the book.

- *Connect receiver to flight controller:* The flight functions such as pitch, yaw, throttle must be connected to each other by pairing appropriate pins. The receiver and the flight controller are placed near each other on the top mounting board of the frame and attached to it. Also, when placing receiver onto frame, it must be kept as far as possible from the power system (motor and battery) to prevent interference. At a minimum it should be kept at a distance of several centimeters from motors or servos.

- *Choose battery and connect to power distribution board:* For a normal hobby drone, you may choose a 3 cell battery. If each cell generates 3.7 V, so it will give a total of 11.1 V. This voltage can later be regulated with the help of voltage regulators or BECs, to different parts of the drone. You must choose a battery that has the same number of connectors as power distribution board. If not, an adapter can be used but this will add to the weight. The power distribution board delivers power to all motors, separately, which all have their own ESCs. The board must be able to handle the total of all peak charges that ESCs draw to transmit to motors. If not, cables can be used, for handling high currents, instead of a PDB. When choosing a battery, you must remember that larger batteries have higher capacities but they are also heavier. Your flying style, weather conditions, weight and center of gravity of the drone, rpm of the motors, and the size and pitch of the propellers are factors that affect battery life. Safety note: when dealing with the battery be careful for not to cause any short circuits and follow specific instructions for battery.

The specifications when choosing battery can include but may not be limited to the items below:

- Minimum Capacity (mAh)
- Voltage
- Number of Cells
- Charge and Discharge Plug
- Constant and Peak Discharge
- Weight
- Dimensions

- *Connect flight controller to frame:* In order to reduce the vibration that the flight controller gets from the frame, vibration isolators such as rubber can be added. The flight controller can be attached to the frame by either sticking it to the frame or using screws.

- *Attach Battery to frame:* The battery, being the heavy component, greatly affects the center of gravity of the frame so its location can be adjusted as needed. The best location for the battery is the center of the drone so that the motors can share the weight equally but of course this depends on the location of other components too.

- *Attach motors to frame:* The motors are first attached to a mounting plate that will hold the motors which would fit to the end of the arms. After attaching these plates to the motors, they should be together connected to the end of each arm.

An assembled drone, with autopilot, ESCs, motors and propeller attached to the frame.
Photo Credit: GetFPV - www.getfpv.com

- *Attach payload:* This is an optional item, in cases where you want to attach a payload to the frame. You must make sure that the payload will not shift during flight and distort the center of gravity of the drone. The attachment points must be firm in order to avoid accidental drops or slips during the flight. Remember that when the drone tilts the most and the payload is the maximum possible, (the worst case) this will create a big turning force (in structural engineering this is called "Moment". Moment = Force x distance) at attachment points when your drone rotates, so do not think that those connection points will only carry a vertical load and make your connections strong enough. Plus there are additional inertial forces due to accelaration and turning of the drone.

- *Attach propellers to motors:* Propellers are attached to motors by adapter rings, which are also called propeller adapters. The adapter ring that comes with the propeller, or you buy it separately but make sure that the inside diameter of the adapter ring fits the motor shaft. The propellers need to be balanced before attaching to motors. Unbalanced propellers will cause vibration for your drone which is bad for many reasons including damaging your motors, loosening of attachment of all components to frame, and cause your photos and videos to be of poor quality. Balancing propeller with the motor at the same time can also be done if necessary, using more advanced techniques. For a quadcopter, prop 1 and 3 should turn in the same direction, say, clockwise (CW), and props 2 and 4 should turn in the opposite counter clockwise (CCW) direction. Choosing the right propeller can make a difference in your drones ease of control and even battery performance so you can try different pitch propellers even for the size you determine, or you may end up with a slightly different size propeller than you considered before with a different pitch or even different number of blades. When buying propellers and motors it is a good idea to order extra pieces. There is a chance that you may break propellers or one of your motors may not work in harmony with the other motors.

- *Attach antenna to frame:* Antenna must be placed as far as possible from cables, carbon fiber and metal parts. It must also be electrically isolated from the frame.

- *Attach the casing to the frame:* Finally the casing should be added as a cover in order to have better aerodynamics as well as protecting the flight controller. Another purpose of casing (shell) is to have an aesthetic appearance. Remember that there are different ways that you can improve the aesthetics of your drone, such as painting the frame and casings, attach led lights, which consume very little power and they weigh very little anyway.

- *Choose Transmitter:* The specifications when choosing the transmitter may include but not limited to the items below:

  - Number of channels
  - Battery Compartment
  - Display Screen
  - Memory
  - Encoder Type

Also make sure that you choose the right transmitter depending on you are right or left handed. Most transmitters have throttle on the left.

A multirotor drone, showing how parts fit together.
Photo Credit: Erle Robotics - www.erlerobotics.com

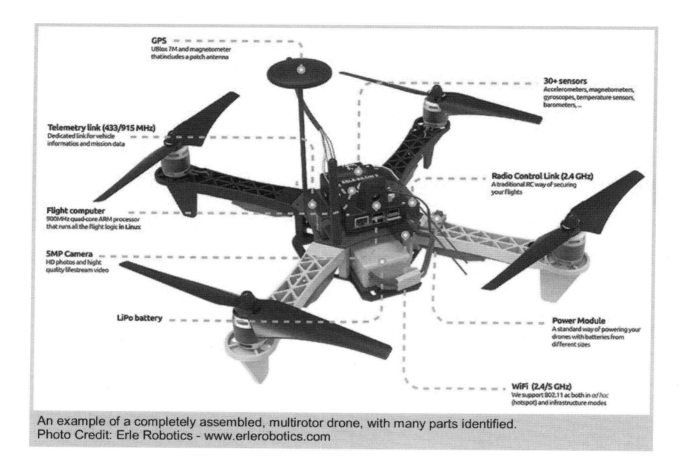

An example of a completely assembled, multirotor drone, with many parts identified.
Photo Credit: Erle Robotics - www.erlerobotics.com

Some Tips when buying drones or drone parts

- If you are buying from a specialty store be careful about the return handling. Make sure that the company will not direct you back to manufacturer country.

- Whether you are buying or building, it is always a good idea to visit one of the forum sites on multirotors or drones and ask advice or at least confirm what you are thinking before you spend money. Even if you are sure, you may be surprised with the answers you get from many experienced people.

- Parts that have higher possibility to break or not function precisely, such as motors or propellers might be ordered in excess to make sure you have spare.

Aerial Photography and Filming - Overview of Important Points

We also want to touch up basics on aerial photo taking and video recording, as most of the time drones are used in filming or photographing purposes.

Discussing cameras, photography and recording videos is a whole other subject by itself and out of scope of this book. An entire book can be written about it. Therefore it is only mentioned here as an overview, particularly in relation to filming with drones and we will leave it at that for now.

Some of the important items to be considered when taking photos or videos with drone:

- The main thing to consider when choosing a camera is to know whether you will mainly take photos or videos and where will you take them.

- There are different types of camera for different purposes. The camera to choose for taking wedding video should be different than a camera needed to take landscape photos, monitoring a ranch or a fast action snowboard video.

- The camera you use must be lightweight and it should be able to stream videos smoothly.

- Frame rate, watertightness, megapixels, dimensions and of course cost are all important factors to consider when choosing.

- The shaking caused by the moving drone must be countered in order to get still images or videos. A gimbal is a good solution as it stabilizes the camera. Also, propellers must be balanced before the flight, in order to reduce shaking caused by unbalanced propellers. For more information, see propeller balancing section under terms and concepts section.

- The light source, sun's position must be considered, just like when you are taking a photo on the ground yourself.

- Taking photos or video during sunrise or sunset hours can create very attractive looking results, as the sun illuminates buildings and natural features differently and the shadows are more defined, adding to the attractiveness of images or videos

- The wind speed increases at higher altitudes, even if it may not be windy at the ground level. Wind causes shaking of the drone, so this must be kept in mind.

- The object of interest must be followed by the camera with proper angle, depending on the position of the drone

- For taking photo or video at night, in addition to headlights, to illuminate the target, colored LED lights should be placed on the drone, so that the operator can see the drone and its orientation

- The photo or video shoot must be planned ahead of time, such as clear line of sight must be maintained always, the timing of video or photo taking must be shorter than the flight time, by taking takeoff and landing times into account

- Rainy and windy weather should be avoided if possible

- The photo of the same target can be taken with different camera settings, so that the optimal mage can be chosen later. This is called Bracketing.

- The photo or video of the target must be captured from different angles, and not from just one angle, to obtain the best result

- Using a Gimbal makes the photo or video quality better, as it stabilizes the camera, but consider the weight of it versus what the drone can carry when buying a gimbal

- The sun's position is important not just for the quality of the video but it may also cause the propellers to cast shadow on to the camera. As a general rule, filming towards the sun should be avoided

- The landing gear must not get in the way of the camera view so it must be chosen accordingly

- Some drones come with non interchangeable cameras, so you must be sure that the camera will serve your purpose, if you are buying such a drone

- FPV, First Person Viewing capabilities of a drone system is also a great factor when choosing a drone for filming or photography

- The speed of the drone should be low, as well as accelerating, decelerating and turning of the drone should also be slow, in order to avoid shaking and get better images or video

- Although being close to target is important, it is also a good practice to increase altitude and capturing more area in the photo or video

- In addition to moving the drone, the gimbal can also be moved at the same time, for more attractive videos.

- The images should be taken in RAW format, which means it has the least minimum processed data from sensors, so that they can be optimally processed later, such as the colors and the exposure of images. A RAW image is an image format that has all the information in it to create an image but it is not yet ready to be printed and needs processing. This gives the ability to optimally process the images later. RAW format images can be considered as digital negatives.

- When capturing photos or videos with a drone, you must make sure that you are not trespassing into private property or violating privacy of others

- Try to move the drone through two axis, such as downwards and forward at the same time, for more professional looking videos. Orbiting around the target and strafing of drone also creates attractive videos

Regulations and Safety Overview

Warning: This section is just an introduction and not an all inclusive list, and it is only written here for the purpose of giving examples and raise awareness about drone safety. You must verify the accuracy of information here and check your particular situation with your drone manufacturer, your neighbors, other experienced drone users, and your local, state and federal authorities as applicable and take all necessary safety precautions about flying drones. Flying a drone must not be viewed as operating a toy, and it requires careful operation and planning by a knowledgeable and skilled operator. Note that some of the items given here can also change at any time so it is important to stay current with all regulations. It is your responsibility to comply with all safety rules and regulations that is applicable to your particular situation. Not observing safety rules for drone operation may cause loss of property or serious injury or even death.

First of all, if you are in the USA, you should be aware of the laws by FAA, Federal Aviation Administration. FAA says that you become part of the US Aviation system, when you fly your drone in nation's airspace. The list below, which is shown in FAA website, gives some basic guidelines:

- Fly below 400 feet and remain clear of surrounding obstacles

- Keep the aircraft within visual line of sight at all times

- Remain well clear of and do not interfere with manned aircraft operations

- Don't fly within 5 miles of an airport unless you contact the airport and control tower before

- Don't fly near people or stadiums

- Don't fly an aircraft that weighs more than 55 lbs

- Don't be careless or reckless with your unmanned aircraft – you could be fined for endangering people or other aircraft

- Drone operators must be at least 13 years old to obtain a certificate

For all other information and more details, permissions, certifications, checklists, reports, we suggest that you carefully check FAA website. There are also many firms that give guidance and provide consultancy services, in order to help on these issues.

Drones also should not be used near wildfires, as they can be dangerous for firefighters, as explained in USDA website.

And here are some other basic safety guidelines (again, this is not an all inclusive list)

- Drones should be operated at a safe distance from vehicles or other people.

- Batteries must be removed after operation.

- Moving parts must be kept clean and dry.

- Drones and drone parts must be kept out of reach of children.

- Moving parts must not be touched.

- Do not operate a drone under influence of alcohol or drugs.

- Wait for a strong GPS signal before takeoff, in order to take advantage of safety features such of returning to base location and position stabilization.

- Drones with damaged parts or wiring must not be operated.

- Do not expose any equipment or parts to water, unless it is specifically designed for it.

- Turn the transmitter on first, before turning on the drone, and after flight, turn off the transmitter last, after turning off the drone.

- Check battery manufacturer's manual and abide by all the safety precautions about using or charging batteries. Remember that improper usage or charging of batteries may not only result in poor performance or falling of your drone, but also can cause fires.

- Make sure you do not have distractions when flying.

- Understand the safety functions of the flight mode that you are using.

- Drones should not be used in snowy, rainy, foggy or too windy weather.

- Prefer open plain areas for flying far from people.

- Make sure that there are no water sand oil or any other foreign objects in the drone.

- Do not talk on telephone when flying a drone.

- Stand at a safe distance from the drone during takeoff.

- Chipped propellers must not be used.

- Do not try to modify motor structure.

- You should use parts that are recommended by your drone manufacturer, and if in doubt, you must contact your manufacturer about the compatibility of a certain part.

- If you think that your drone is definitely about to crash, turn the throttle to zero.

- Flying near broadcasting towers, high voltage lines is not safe, as these affect magnetic and radio waves.

- Remove all batteries from the drone, before doing any work on it. If for some reason you must keep the battery there, then remove all the propellers before doing any work. The propellers of a drone can damage your hand badly.

- Initiate timer when starting flight, if it is not automatic.

- Do not trespass private property.

- The drone must be landed as soon as possible, if it starts to drift excessively. Chances are the compass is malfunctioning.

- Start flying with full batteries. Even if the indicator may show partially full battery, it may not be accurate and you may have less than expected flight time, when you start with a partially full battery.

- The higher you fly, the more careful you must be. It also gets windier at higher altitudes.

- When recording with camera, respect privacy of others.

- Read and understand all safety instructions for your drone and your particular situation.

Drone operators should also log their flights into a flight log to keep records of the flight such as date time location etc... Check the official requirements about this. There are even online services that help doing this online.

The rules and regulations on drones are still evolving as this is a new area. As of end of 2015, FAA now requires all drones which weigh over 0.55 pound in the United States to be registered until February 2016. The drones operated indoors only however, do not need registration. You can also check the FAQ section on FAA website at: https://www.faa.gov/uas/registration/faqs/

Companies in Drone Business

Note: We have no affiliation with or endorse any of the companies or websites below, including the ones that has introductory banners. **The text in the banners belong to respective companies,** and we did not verify those texts. The companies below are the ones we could identify as of the date of writing. You can find both hobby grade or advanced grade drone manufacturers, parts manufacturers, seller and service companies here. If you think your company should be here, please contact us.

Argentina

Aerovision - http://www.aero-vision.com.ar/ - drone services

Hobby10 - http://www.hobbydiez.com.ar - seller

Australia

Aerobot.com - http://www.aerobot.com.au/ - drone services

Aerosonde - http://www.aerosonde.com - manufacturer, advanced systems

Codarra - http://www.codarra.com.au/ - manufacturer, consulting, drone services

HEI - http://www.hobbyexpress.com.au/ - wholesale

Riseabove - http://www.riseabove.com.au - manufacturer, seller, drone services

Silvertone - http://www.silvertone.com.au- manufacturer, advanced systems

Silvertone Electronics is an Australian owned and operated design and manufacturing organization of long range, high endurance, high lift UAVs for the commercial market. Committed to safe skies for all our quality system and manufacturing processes ensure a reliable platform for all commercial operations. Silvertone works closely with the local NAA (CASA) in developing the standards for UAV design and manufacture and is working with industry to open Beyond Visual Line of Sight operations.
The Silvertone team boasts more than 70 years combined aviation experience both as operators and in the field of aircraft design and certification. We are driven to provide a UAV platform with operational support that exceeds current regulations.

Skydrone - http://www.skydrone.com.au/ - drone services

UnmannedSystems - http://www.unmannedsystemsaustralia.com.au/ -manufacturer,advanced systems

V-Tol Aerospace - http://v-tol.com/ - manufacturer, advanced systems

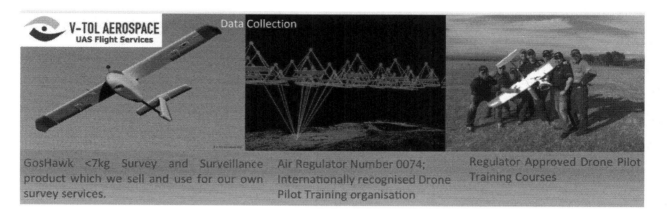

GosHawk <7kg Survey and Surveillance product which we sell and use for our own survey services.

Air Regulator Number 0074; Internationally recognised Drone Pilot Training organisation

Regulator Approved Drone Pilot Training Courses

XM2- http://www.xm2.com/ - drone services

Austria

Airborne Robotics -http://www.airborne-robotics.com/ - manufacturer

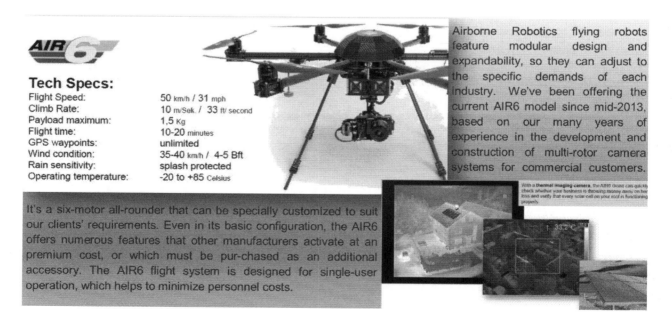

Tech Specs:

Flight Speed: 50 km/h / 31 mph
Climb Rate: 10 m/Sek. / 33 ft/ second
Payload maximum: 1,5 Kg
Flight time: 10-20 minutes
GPS waypoints: unlimited
Wind condition: 35-40 km/h / 4-5 Bft
Rain sensitivity: splash protected
Operating temperature: -20 to +85 Celsius

Airborne Robotics flying robots feature modular design and expandability, so they can adjust to the specific demands of each industry. We've been offering the current AIR6 model since mid-2013, based on our many years of experience in the development and construction of multi-rotor camera systems for commercial customers.

It's a six-motor all-rounder that can be specially customized to suit our clients' requirements. Even in its basic configuration, the AIR6 offers numerous features that other manufacturers activate at an premium cost, or which must be pur-chased as an additional accessory. The AIR6 flight system is designed for single-user operation, which helps to minimize personnel costs.

Dynamic Perspective - http://dynamicperspective.com/home/ - manufacturer

Flitework - http://en.flitework.at/ - manufacturer

Lindinger - http://www.lindinger.at - seller

Robitronic - http://www.robitronic.com/en/ - seller

Shiebel - http://www.schiebel.net/ - manufacturer, advanced systems

Belarus

Indela - http://www.indelauav.com/ - manufacturer, advanced systems

"INDELA–SKY" Unmanned Aerial System is designated for the traditional monitoring functions performance, actions coordination and surveillance.
1. Real time terrain surveillance, all-time, all-climate;
2. Information transmission and processing in visual and infrared specters;
3. Target search and acquisition, target video capture and position determination;
System INDELA–SKY is a fully autonomous and self-sufficient: the UAV is deployed from transport condition to the takeoff readiness within 15 min.

INDELA-I.N.SKY is a medium range unmanned aerial vehicle (UAV) and designed for day/night remote monitoring and transmission of video information to the ground control station in real time. The operation range is up to 100 km in line of sight from the ground point; operation time is up to 5 hours. Maximum take-off weight is 140 kg. The payload weight is up to 25 kg.

Belgium

Aerobertics - http://www.aerobertics.be/ - seller

Altigator - http://altigator.com/en/ - manufacturer, seller, drone services

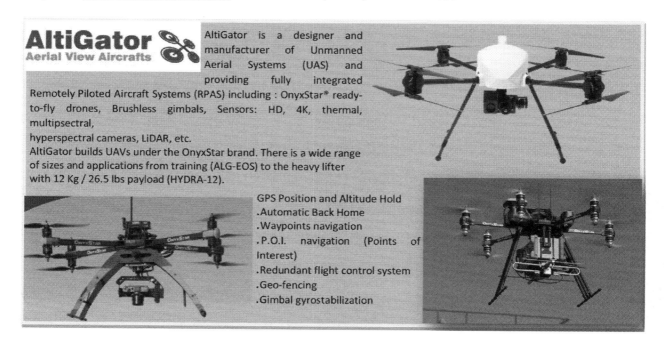

AltiGator is a designer and manufacturer of Unmanned Aerial Systems (UAS) and providing fully integrated Remotely Piloted Aircraft Systems (RPAS) including : OnyxStar® ready-to-fly drones, Brushless gimbals, Sensors: HD, 4K, thermal, multipsectral, hyperspectral cameras, LiDAR, etc.
AltiGator builds UAVs under the OnyxStar brand. There is a wide range of sizes and applications from training (ALG-EOS) to the heavy lifter with 12 Kg / 26.5 lbs payload (HYDRA-12).

GPS Position and Altitude Hold
.Automatic Back Home
.Waypoints navigation
.P.O.I. navigation (Points of Interest)
.Redundant flight control system
.Geo-fencing
.Gimbal gyrostabilization

Brazil

Aerofoundry - http://www.aerofoundry.com/ - manufacturer, advanced systems

Aileron Modelismo - http://aileronmodelismo.com.br - seller

FT Sistemas - http://flighttech.com.br/ - manufacturer, advanced systems

Gohobby Distribuidora - http://www.gohobby.com.br/DJI- seller

Gyrofly - http://www.gyrofly.com.br/en/ - manufacturers, consulting, drone services

Canada

Aerial Tech - http://aerialtech.com/ - seller, drone services, wholesale

Aeromao - http://www.aeromao.com/ - manufacturer, seller, drone services

Aeromao Inc. manufactures the Aeromapper series of unmanned aerial vehicles destined for image aqcisition for mapping and surveying applications. The Aeromappers are currently operating in more than 40 countries by many clients including government institutions, private companies, universities and research organizations in a wide variety of applications and tasks. Our drones easily outperform most of the survey drones out there in the market.

The Aeromappers offer the best combination of optics, multisensors capabilities via quick swappable mounts, parachute recovery, strength, reliabilty and ease of use. Each drone we manufacture is individually flight tested and delivered with a detailed step by step manual for users with cero experience. The Aeromappers are turnkey systems, tested and delivered ready to use with no assembly or setup required. Start acquiring data right away

Aeryon Labs - http://aeryon.com/ - manufacturer, advanced systems

Alphadrones - http://www.alphadrones.ca/ - seller

Analytic Systems - http://www.analyticsystems.com/ - parts manufacturer

Applanix -http://www.applanix.com/ - seller

Challis Heliplane - http://www.challis-heliplane.com - manufacturer, advanced systems

Drone Shop Canada - https://www.droneshopcanada.ca/ - seller, repairs

Droneology - http://www.droneology.ca/ - seller

Flying Cameras - http://store.flyingcameras.com/ - seller

Great Hobbies - https://www.greathobbies.com/ - seller

Heli Video Pros - http://helivideopros.com/wp/ - seller, drone services

Hobby Hobby - http://hobbyhobby.com/store/ - seller

Ingrobotic - http://ingrobotic.com/ - manufacturer, drone services

Multidrone - http://multidrone.com/ - seller, drone services

Multirotor Heli - http://www.multirotorheli.ca/ - seller

Novatel - http://www.novatel.com/ - part manufacturer

Propphotouav - https://www.prophotouav.com/ - seller

Stratus Aeronautics - http://www.stratusaeronautics.com/ - manufacturer, advanced systems

Chile

Drone Center - http://www.dronecenter.cl/ - seller

China

Alltech - http://en.keweitai.com/en/index.aspx - manufacturer advanced systems

Art Tech - http://www.art-tech.com/en/ - manufacturer, seller

Diatone - http://www.diatone.hk - manufacturer

DJI - http://www.dji.com/ - manufacturer

Dongguan Hiee Electronic - http://www.fpvhiee.com/ - parts manufacturer

Dualsky - http://www.dualsky.com/ - manufacturer

DYS - http://www.dys.hk/ - electronic parts manufacturer

Ehang - http://www.ehang.com/ - manufacturer

Ehirobo - http://www.ehirobo.com/ - seller

Emlid - http://www.emlid.com - manufacturer, seller

EMLID Emlid is a fast growing company that was first to introduce an autopilot system based on Raspberry Pi - Navio. Autopilot gained popularity for its unrivalled processing power, ease of programming and performance previously unseen in microcontroller based systems. To improve navigation precision Emlid developed a low cost and easy to use RTK system Reach. It finds application in drones, agriculture and surveying. With revolutionary price point Emlid is bringing high accuracy RTK positioning to new markets and industries.

Ev-peak - http://www.ev-peak.com/ - manufacturer

Feishen Group Company - http://www.fs-racingart.com/uav.php - manufacturer

Flypro Aerospace Tech - http://www.flypro.com/en - manufacturer

FLYPRO is an innovative company specializing in developing and innovating Unmanned Aerial Vehicle (UAV). Our internationally team produce industry-leading designs involving advanced flight control systems, visual positioning systems and obstacle avoidance.

FLYPRO designs for sports, focusing on an intelligent simple control mode, the ultimate fun user experience and cost-effective innovative products. Independent design and development of the world's first automatic UAV which is the only UAV with a smart watch follow PX400 FPV UAV and the XEagel control system and automatic obstacle avoidance. FLYPRO has a number of technological innovations and functional features to fill gaps in the industry. FLYPRO has established a benchmark position in the market of intelligent motion UAV market segment.FLYPRO's focus is on the people who like outdoor activities; we are devoted to developing creative smart sports UAV.

FR Sky - http://www.frsky-rc.com/ - electronic parts manufacturer

Gemfan - http://www.gemfanhobby.com/En - parts manufacturer

Goodluckbuy - http://www.goodluckbuy.com - seller

Guandong Cheerson - http://www.cheersonhobby.com/ - manufacturer

CHEERSON

Cheerson Hobby is a professional remote control toys and drones manufacturer which integrates researching, manufacturing and marketing, and it has its own more than 13,000 square meters factory. With over ten years rapid development, Cheerson Hobby's self-developed remote control toys and drones have received recognition and praise in its industry. CXHOBBY brand and SH brand also got high honor in the world.

Cheerson Hobby has been continuously consolidating its management system and has got ISO9001 Management System Certification, QAC Quality Management System Certification, and a number of international certifications such as ICTI, FCCA, GSV, SCS, CE(EN71,EN62115,R&TTE) ROHS,FCC,ASTM F-963, etc. It has also been approved to obtain the registration certificate for export toy products. Products are exported to all over the world.

Helipal - http://www.helipal.com/ - seller

Hj-Toys - http://www.hj-toys.com/ - manufacturer

Hobby - Wing - http://www.hobby-wing.com/index.html - seller

Hobbywing - http://www.hobbywing.com/ - electronic parts manufacturer

Hubsan - http://www.hubsan.com/ - manufacturer

We are manufacturer of RC drones. We have our own R&D group and also have factory to produce our products.

H501S
HUBSAN X4 FPV BRUSHLESS

Idea-Fly - http://www.idea-fly.com/ - manufacturer

Immersion RC - http://www.immersionrc.com/ - manufacturer, seller

Keweitai - http://en.keweitai.com/ - manufacturer

Makerfire - http://www.makerfire.com/ - manufacturer

Maytech - http://www.maytech.cn/en/ - parts manufacturer

Maytech Electronics Co., Ltd.designs, manufactures carbon fiber propellers, brushless Electronic Speed Controllers (ESCs), motors for UAV, r/c models. Carbon props size from 3inch to 30inch, with pure or omposite carbon fiber materials. All props are pre-balanced in production line. Hole type can be customized. For ESCs, Maytech now introduces out new line of ESCs with EFM8BB10F8 MCU AND 32bit ARM Micreoprocesser. Small Size for smaller racing quads. Enjoy a more efficient, smoother, more responsive flight with a powerful 32-bit MCU for super fast clock rates. Brushless motor features of high efficiency, low-heat,good stability. Mainly include brushless Gimbal motors, multikopter motors, Outrunner motors, car motors. Maytech products are popularly distributed in Europe, America, Asia, Australia and other regions. Quality and service win customers praise worldwide.

MJX - http://www.mjxtoys.com/ - manufacturer

Nine Eagles - http://www.nineeagle.com/ - manufacturer

Small Light UAV with 3 Axis Gimbal and high—def camera
Control by smartphone
Automatic return
Lightweight with small size
High quality with reasonable price
Folding wings for easy carry
Capture from multi angles by one touch
Intelligent flying selfie mode
Automatic flying to designated height
Real time video / image transmission

Breeze sharing via social networks
Intelligent gimbal lifting system
One key return home, landing,
One key altitude hold, take-off
One key fixing flight distance
Automatically memorize departure point
Return home button
Intelligent route planning
Follow me flight
Point to point autonomous flight

Orbit flight
Memorize flight routes Emergency Landing
Auto Avoidance of no fly zones

RC Logger - http://www.rclogger.com/ - manufacturer

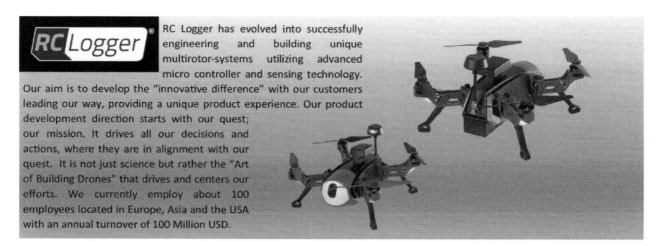

RC Logger has evolved into successfully engineering and building unique multirotor-systems utilizing advanced micro controller and sensing technology. Our aim is to develop the "innovative difference" with our customers leading our way, providing a unique product experience. Our product development direction starts with our quest; our mission. It drives all our decisions and actions, where they are in alignment with our quest. It is not just science but rather the "Art of Building Drones" that drives and centers our efforts. We currently employ about 100 employees located in Europe, Asia and the USA with an annual turnover of 100 Million USD.

Radiolink - http://www.radiolink.com.cn/doce/index.html - Transmitter, receiver manufacturer

RC Mart - http://www.rcmart.com - seller

Skyartec - http://www.skyartec.com/ - manufacturer

Shenzen AEE - http://www.aee.com - manufacturer

Sky RC - http://www.skyrc.com/ - manufacturer

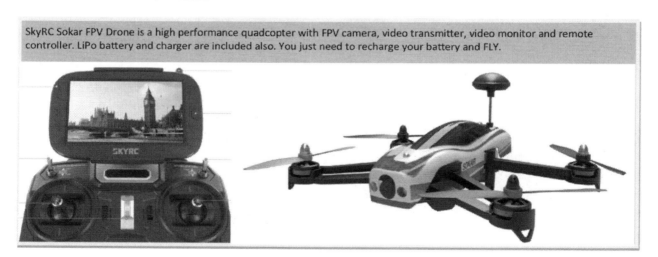

SkyRC Sokar FPV Drone is a high performance quadcopter with FPV camera, video transmitter, video monitor and remote controller. LiPo battery and charger are included also. You just need to recharge your battery and FLY.

Sunnysky Motors - http://www.rcsunnysky.com/ - motor manufacturer

Symatoys - http://www.symatoys.com/ - manufacturer

T Motor - http://www.rctigermotor.com/ - parts manufacturer

Tarot RC - http://www.tarotrc.com - manufacturer

Walkera - http://www.walkera.com/ - manufacturer

XY Aviation - http://www.xy-aviation.com/en/index.aspx - manufacturer

Yuneec - http://www.yuneec.com/ - manufacturer

Colombia

Air Photo Colombia - http://www.airphotocolombia.com/ - seller, drone services

Compudemano - https://www.compudemano.com/ - seller, repairs

Heliboss Colombia - http://heliboss.co/ - seller, drone services, repairs

Czech Republic

Aerodron - https://www.aerodron.ch/cms/ - drone services

Axi Model Motors - https://www.modelmotors.cz/ - motor manufacturer

Hacker - http://www.zoomport.eu/shop/ - seller

Jetimodel - http://www.jetimodel.com/en/ - electronic parts manufacturer

MS Composit - http://www.mscomposit.info/ - manufacturer

Pelikan Daniel - http://www.pelikandaniel.com/ - seller

Pelikan Drone - http://www.drona.cz/ - seller

Denmark

Dandrone - http://dandrone.dk/ - seller, drone services

Danish Aviation Systems - http://www.danishaviationsystems.dk/ - manufacturer, advanced systems

North TQ - http://www.northtq.com/ - seller

Sky Watch - http://sky-watch.dk/ - manufacturer, advanced systems

Sky-Watch has since 2009 developed, manufactured and implemented high-tech solutions for real-time decision making in complex environments worldwide. Sky-Watch's R&D competencies encompasses unmanned systems (primarily UAV technology for professionals), advanced embedded control software, integrated industrial design and intuitive user interfaces

- Lightweight (1.7 kg)
- Small and Portable
- Easy-to–use navigator
- Ruggedized & Waterproof
- 2 hours flight time
- Precise navigation system
- Hand-Launched

- Accuracy Deep-Stall landing (10x10 m)
- Real time night & day video feed
- Encrypted duplex communication

- Lightweight
- Small and Portable
- Small Footprint
- Easy-to–use Navigator
- Ruggedized & Waterproof
- 25 min flight time
- Drag & Drop Navigation system
- Hand-Launched
- Real time night & day video feed
- Encrypted communication

Estonia

Threod Systems - http://www.threod.com/ - manufacturer, advanced systems

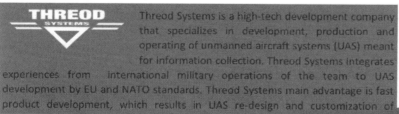

Threod Systems is a high-tech development company that specializes in development, production and operating of unmanned aircraft systems (UAS) meant for information collection. Threod Systems integrates experiences from international military operations of the team to UAS development by EU and NATO standards. Threod Systems main advantage is fast product development, which results in UAS re-design and customization of standard solutions within several months. Thus, the systems can be quickly adapted to operating environment, which is crucial for supporting combat operations or other information collection activities.

Finland

Aerotekniikka - http://www.aerotekniikka.fi/ - manufacturer, advanced systems

Videodrone - http://www.videodrone.fi/ - manufacturer

GeoDrone® is Finnish designed and manufactured high-end Professional Aerial Mapping System. GeoDrone® systems are always supplied as a complete package. The package includes the aircraft, camera, radio control unit, batteries, charger, aluminium transport case and flight planning tool.

France

Axion Drone - http://axiondrone.com/ - drone services

Delair Tech - http://www.delair-tech.com/en/home/ - manufacturer

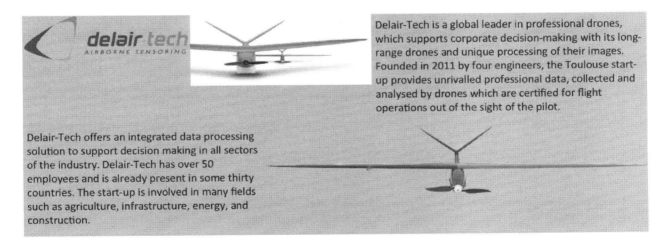

Delair-Tech is a global leader in professional drones, which supports corporate decision-making with its long-range drones and unique processing of their images. Founded in 2011 by four engineers, the Toulouse start-up provides unrivalled professional data, collected and analysed by drones which are certified for flight operations out of the sight of the pilot.

Delair-Tech offers an integrated data processing solution to support decision making in all sectors of the industry. Delair-Tech has over 50 employees and is already present in some thirty countries. The start-up is involved in many fields such as agriculture, infrastructure, energy, and construction.

Dronexclusive - http://www.dronexclusive.com/ - seller

Drone Shop - http://www.droneshop.com/ - seller, drone services

Drone Volt - http://www.dronevolt.com/ - drone services

Flash RC - http://www.flashrc.com/ - seller

Flying Robots - http://www.flying-robots.com/en/ - manufacturer, advanced systems

Fly-n-Sense - http://www.fly-n-sense.com/ - manufacturer, advanced systems

Fnac - http://www.fnac.com/ - seller

Helipse - http://www.helipse.com/ - manufacturer, advanced systems

Novadem - http://www.novadem.com/ - manufacturer

RC Team - http://www.rcteam.fr/ - seller

SBG Systems - http://www.sbg-systems.com/ - inertial sensors manufacturer

 SBG Systems offers a complete line of inertial sensors based on the state-of-the-art MEMS technology such as Attitude and Heading Reference System (AHRS), Inertial Measurement Unit (IMU), Inertial Navigation Systems with embedded GPS (INS/GPS), and more... Our sensors are ideal for industrial, defense & research projects such as unmanned vehicle control, antenna tracking, camera stabilization, and surveying applications.

Small, light-weight and accurate, our miniature inertial navigation sensors are ideal for unmanned systems autonomous navigation and orientation, wether they are Unmanned Air Vehicles (UAV), Unmanned Ground Vehicle (UGV) or Unmanned Marine Vehicles (UMV). Ellipse Series is a brand new product family of miniature inertial sensors. It provides roll, pitch, heading, heave, and position data in real-time. Ellipse Series comes with features inspired from high end inertial systems such as FIR and anti-rejection filtering, robust IP68 enclosure, high output rate, RTK corrections, automatic alignment, and more...

Survey Copter - http://www.survey-copter.com/ - manufacturer

Xamen - http://www.xamen.fr/index.php/fr/ - manufacturer

Pilot Station

Carrying Case

Xamen Technologies Xamen Technologies is an innovative French company which designs and develops unmanned aircraft in strict compliance with the industry legislation.
We have developed **the first ATEX Certified UAV in the world**. This exclusive innovation is capable to carry specific sensors (Camera, gas detectors, concentration sensors, etc...) to unreachable zone in an ATEX environment. This solution provides the most secure way for your maintenance and security requirement, without suspension of usual operation and production.

The focus has been put on safety and all associated services such as training and maintenance. All our aircraft feature GPS-assisted flight, failsafe return to base and parachute (impact of maximum 69 joules). Our 180 Degrees gimbal tilt angle, from facing over 90 straight down to 90 straight up, is ideal for inspection work.

Germany

Aerolab - https://www.aerolab.de/ - manufacturer, seller

Aeronaut - http://www.aero-naut.de/en/home/ - manufacturer, seller

AF Marcotec - http://www.marcotec-shop.de- seller, repairs

Aibotix - https://www.aibotix.com/en/ - manufacturer

Airobot - http://www.airrobot.com/ - manufacturer

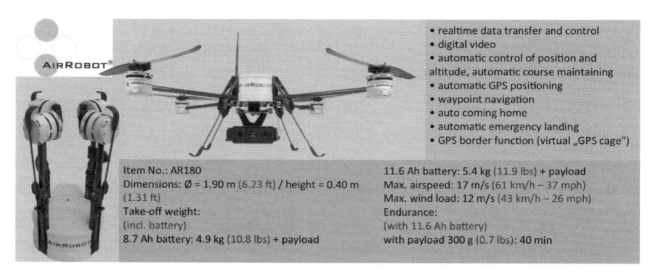

Item No.: AR180
Dimensions: Ø = 1.90 m (6.23 ft) / height = 0.40 m (1.31 ft)
Take-off weight:
(incl. battery)
8.7 Ah battery: 4.9 kg (10.8 lbs) + payload

11.6 Ah battery: 5.4 kg (11.9 lbs) + payload
Max. airspeed: 17 m/s (61 km/h – 37 mph)
Max. wind load: 12 m/s (43 km/h – 26 mph)
Endurance:
(with 11.6 Ah battery)
with payload 300 g (0.7 lbs): 40 min

- realtime data transfer and control
- digital video
- automatic control of position and altitude, automatic course maintaining
- automatic GPS positioning
- waypoint navigation
- auto coming home
- automatic emergency landing
- GPS border function (virtual „GPS cage")

Alexander Engel KG - http://www.engel-modellbau.eu/ - manufacturer, seller

Ascending Technologies - http://www.asctec.de/ - drone services

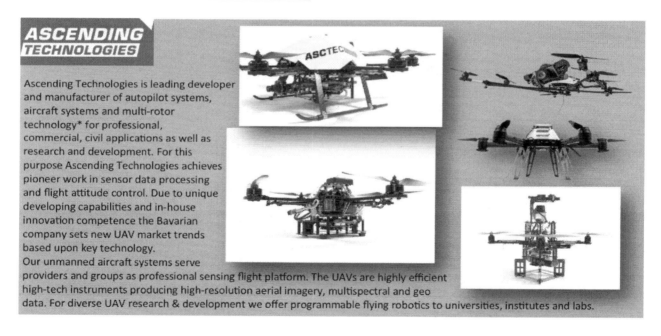

ASCENDING TECHNOLOGIES

Ascending Technologies is leading developer and manufacturer of autopilot systems, aircraft systems and multi-rotor technology* for professional, commercial, civil applications as well as research and development. For this purpose Ascending Technologies achieves pioneer work in sensor data processing and flight attitude control. Due to unique developing capabilities and in-house innovation competence the Bavarian company sets new UAV market trends based upon key technology.
Our unmanned aircraft systems serve providers and groups as professional sensing flight platform. The UAVs are highly efficient high-tech instruments producing high-resolution aerial imagery, multispectral and geo data. For diverse UAV research & development we offer programmable flying robotics to universities, institutes and labs.

Aviotiger Germany - https://www.aviotiger-germany.de/ - seller

Bavarian Demon - http://www.bavariandemon.com/en/ - electronic parts manufacturer

Birdpilot - http://www.birdpilot.com/en/home - manufacturer

Braeckman Modellbau - http://www.braeckman.de/ - manufacturer

Captron - http://www.captron.de/en/ - sensor manufacturer

D - Power - http://www.d-power-modellbau.com/ - manufacturer, seller

Decker Planes - http://www.decker-planes.de/ - manufacturer

Donau Elektronik - http://www.donau-elektronik.com/ - parts manufacturer

Droneparts - http://www.droneparts.de/ - seller

Dryfluids - http://www.dry-fluids.com/dryfluid-heli-1.html - parts manufacturer

Emcotec - http://www.emcotec.de/ - embedded controllers manufacturer

EMT Penzberg - http://www.emt-penzberg.de/home.html - seller

Flyduino - http://www.flyduino.net/ - seller

FPV24 - http://www.fpv24.com- seller

Freakware - https://www.freakware.de- seller

Gensace - http://www.gensace.de/ - battery manufacturer

Globe Flight - http://www.globe-flight.de/ - seller, drone services

Graupner - http://www.graupner.de - manufacturer, seller

Hoelleinshop - http://www.hoelleinshop.com - seller

IRC Electronic - http://shop.rc-electronic.com/ - manufacturer, seller

Kontronik Drives - http://www.kontronik.com - motor and electronic components manufacturer

LF-Technik - http://www.lf-technik.de/ - manufacturer

Microdrones - https://www.microdrones.com/en/home/ - manufacturer, drone services

Mikado Model Helicopters - http://shop.mikado-heli.de/ - manufacturer, seller

Mikocopter - http://www.mikrokopter.de/en/home - manufacturer, drone services

Miniprop - http://www.miniprop.com/ - manufacturer

Modellbau -http://www.mhm-modellbau.de/ - seller

MTTec - http://www.mttec.de/ - seller

Multiplex - http://www.multiplex-rc.de/en.html - manufacturer

Optotronix - http://www.optotronix.de/ - manufacturer, advanced systems

Plettenberg - http://www.plettenberg-motoren.net/ - motor manufacturer

Powerbox Systems - https://www.powerbox-systems.com/ - electronic and power parts manufacturer

Premium Modellbau - http://www.premium-modellbau.de/ - seller

RC Hub - http://www.rc-hub.com - seller

Rosewhite - http://www.rosewhite.de/ - manufacturer

Service Drone - https://www.service-drone.com/en/ - manufacturer, drone services

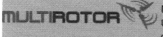 

MULTIROTOR is a pioneer in the UAV market. Long before UAV were commonly known and recognized MULTIROTOR has developed the first micro-multicopter for future-oriented B2B applications in 2008.

MULTIROTOR has been the pioneer of the leading technology trends over the last years. Our company can present the biggest experience and customer base in the market of professional multicopter drones for commercial application.

The fast growing company produces series of individual flight drones as end-to-end solutions for an increasing number of astonishing applications. The production and assembly work of all key components, such as hard- and software, are 'made in Germany' and of the highest quality. Multirotor is producing systems which even meet the security standards oriented towards the ones of civil aviation. The high scalability enables an almost unlimited range of applications and operations.

SLS Stefans LipoShop - http://www.stefansliposhop.de/liposhop/ - battery manufacturer

SM Modellbau - https://www.sm-modellbau.de/ - parts manufacturer

Sobek Drives - http://sobek-drives.de/ - motor and electronic components manufacturer

The Missing Gear - http://www.themissinggear.eu/ - camera manufacturer

Greece

Multicopter - http://multicopter.gr/en/ - seller

My Helis - http://www.myhelis.com/ - seller

Hungary

Extreme Digital - http://www.edigital.hu/ - seller

Modell & Hobby - http://shop.modell.hu/ - seller

India

Ascom - http://www.flycams.in/products.html - seller

Aurora - http://www.aurora-is.com/ - manufacturer, advanced systems

Kadet Defence Systems - http://www.kadet-uav.com/ - manufactuer, advanced systems

Om UAV Systems - http://www.omuavsystems.com/ - manufacturer, seller

RC Mumbai - http://rcmumbai.com/ - seller

Robu.in - http://www.robu.in - seller

Indonesia

Aero Terrascan - http://www.aeroterrascan.com/ - manufacturer, drone services

Ireland

Model Heli Services - http://www.modelheliservices.com/ - seller

Israel

Bluebird Aero Systems - http://www.bluebird-uav.com/ - manufacturer, advanced systems

Contorp - http://www.controp.com/ - manufacturer, advanced systems

Innocon - http://www.innoconltd.com/ - manufacturer, advanced systems

Steadicopter - http://www.steadicopter.com/ - manufacturer, advanced systems

Italy

Aermatica - http://www.aermatica.com/ - manufacturer, advanced systems

Bizmodel - http://www.bizmodel.it/ - seller

Drone World SRL - http://www.droniworld.com/ - manufacturer

HobbyHobby - http://www.hobbyhobby.it/ - seller

Jonathan - http://www.jonathan.it/ - seller

Mavtech - http://www.mavtech.eu/ - manufacturer, drone services

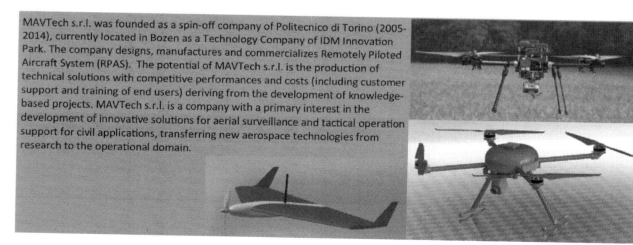

MAVTech s.r.l. was founded as a spin-off company of Politecnico di Torino (2005-2014), currently located in Bozen as a Technology Company of IDM Innovation Park. The company designs, manufactures and commercializes Remotely Piloted Aircraft System (RPAS). The potential of MAVTech s.r.l. is the production of technical solutions with competitive performances and costs (including customer support and training of end users) deriving from the development of knowledge-based projects. MAVTech s.r.l. is a company with a primary interest in the development of innovative solutions for aerial surveillance and tactical operation support for civil applications, transferring new aerospace technologies from research to the operational domain.

MS Heli - http://www.msheli.com/ - manufacturer

Sab Heli Division - http://www.sabitaly.it/ - manufacturer

Japan

Drone Express - http://www.droneex.net/en/ - manufacturer

Futaba - http://www.futabarc.com/ - parts manufacturer

Kazakhstan

Ruba Technology - http://wifi.kz/ - seller

Korea

UMAC Air - http://www.umacair.com/ - manufacturer

Gryphon Dynamics - http://gryphondynamics.co.kr/ - manufacturer

Secraft - http://www.secraft.net/index.php - seller

Latvia

Drone Technology - http://www.dronetechnology.eu/ - manufacturer

Lithuania

Elektromarkt - http://www.elektromarkt.lt/ - seller

Promaksa - http://www.promaksa.lt/ - seller

Malaysia

GKS Hobby - http://gkshobby.com/ocart/ - seller

Mexico

Dronics - http://www.dronics.mx/ - seller

Heliboss - http://www.heliboss.com/ - seller, repairs, drone services

Quad Drone Base - http://www.quaddronebase.com/ - seller, drone services

Netherlands

Ace Core Technologies - http://www.acecoretechnologies.com/ - manufacturer

Aerialtronics - http://www.aerialtronics.com/ - manufacturer

Aerialtronics revolutionizes the way business resources are managed with our automated B2B compatible drones. We provide aerial data solutions for a range of commercial applications worldwide.

The Altura Zenith is designed and developed according to aviation grade quality standards.

Birds-Eye-View - https://www.birds-eye-view.nl/ - manufacturer

Clear Flight Solutions - http://www.clearflightsolutions.com/ - manufacturer

Clear Flight Solutions is a young company, specializing in two fields: bird control and (industrial) inspections. With the combined use of our unique Robirds, drones and other exciting new technologies, we offer an unmatched level of effectiveness in both fields. Birds are beautiful creatures. However, if you work in aviation, harbours, waste management or agriculture, you will be aware that birds can be a very tough problem to deal with. Besides being a nuisance, birds can also form a serious threat to safety in aviation. The Robird is an environmentally-friendly solution for all your bird-related problems. Robirds are truly unique remotely controlled robotic birds of prey, with the realistic appearance and weight of their living counterparts. Combined with other proven techniques, the Robird is the most effective way of bird control.

Delft Dynamics - http://www.delftdynamics.nl/index.php/en/ - manufacturer

Droneland - http://www.droneland.nl/en/ - seller

High Eye - http://www.higheye.nl/company/ - manufacturer

New Zeland

Aeronavics - http://aeronavics.com/ - manufacturer

Hawk Uas - http://www.hawkuas.com/ - manufacturer, drone services

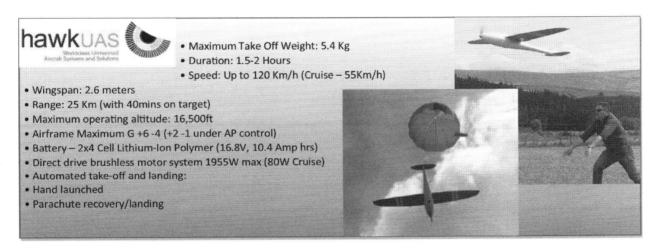

Skycam UAV - http://www.kahunet.co.nz/ - manufacturer, drone services

Norway

Elefun - http://www.elefun.no - seller

Proxy Dynamics - http://www.proxdynamics.com/home - manufacturer, advanced systems

RC Butikken - http://rcbutikken.no/ - seller

Robot Aviation - http://www.robotaviation.com/ - manufacturer, advanced systems

Simicon - http://simicon.no/ - parts manunfacturer

Pakistan

New Digital City - http://www.newdigitalcity.com/ - seller

Integrated Dynamics - http://www.idaerospace.com/ - manufacturer, advanced systems

Satuma - http://www.satuma.com.pk/ - manufacturer, advanced systems

Paraguay

Gabahobby - http://www.gabahobby.com/ - seller

Philippines

RC Victory World - http://www.rcvictory.com/ - seller

Poland

Dilectro - http://dilectro.pl/ - seller

Multirotor - http://www.multirotor.co/ - seller

Portugal

Aeroazores - http://www.aeroazores.com/ - manunfacturer, drone services

Divercentro - http://diver.pt- seller

HP Modelismo - http://www.hpmodelismo.com/ - seller

Timeblocks - https://timeblocks.pt/pt/ - seller

Uavision - http://www.uavision.com/ - manufacturer

Puerto Rico

Techys - http://www.techysdrones.com/ - seller, consulting

Romania

Aft - http://www.aft.ro/ - manufacturer, advanced systems

Sierra Modellsport - http://www.sierra.ro/ - seller

Russia

Bratya Ryat - http://brrc.ru/ - seller

Enics - http://www.en.enics.ru/ - manufacturer, advanced systems

RC Copter - http://rccopter.ru/ - seller

RC Team - http://www.rcteam.ru/ - seller

Skymec - http://skymec.ru/ - seller, repairs

Turborobot - http://turborobot.ru/ - seller, repairs

San Marino

Sebart - http://www.sebart.it/ - manufacturer

Singapore

Flying Bots PTE - http://flyingbots.net/ - seller, drone services

Oqualia - http://www.oqualia.com/ - manufacturer, drone services

The Drone Shop - https://thedroneshop.co/ - seller

Slovakia

RC Svet - http://www.rcsvet.sk/ - seller

Xtreme - http://www.xtreme.sk/ - seller

Slovenia

C-Astral - http://www.c-astral.com/ - manufacturer, advanced systems

C-Astral is an aerospace enterprise and solution provider based in Ajdovscina, Slovenia, the "hub" of advanced aerospace development and integration in this part of the world. The company is one of the market leaders in the small unmanned systems (UAS) and services field and has a global presence, a robust research and development program and advanced integration/customization capacities. The company is built around the fields of expertise and practical experience in aerospace, unmanned systems, electronics and sensor

development, aerial based surveying and processing, remote sensing, telecommunications, renewable energy systems and extreme environment autonomous habitats and communications. C-Astral operates a software and hardware laboratory for aerodynamics and systems integration work and a prototyping CAD/

CAM workshop facility for composite and metal materials work, modeling and systems integration.

South Africa

Aerial Monitoring Solutions - http://www.ams.za.com/ - manufacturer, advanced systems

Making use of a high efficiency blended-wing-body design, the Eagle-Owl system offers the customer an unparalleled multi-purpose aerial platform. The modular design of the aerial platform, allows for easy transportation, individual customization as well as reduction of downtime during maintenance or repairs.

Aerial Monitoring Solutions identifies the needs of the customer and offers the complete aerial platform solution including:
o Airframe
o Engine
o Auto-pilot System
o On-board cameras
o Launch and retrieval systems
o Ground Station and telemetry
o Training
o Maintenance

Cat UAV - http://www.catuav.com/ - manufacturer, drone services

Embention - http://www.embention.com/en/home.htm - manufacturer, advanced systems

Spain

Alpha Unmanned Systems - http://www.alphaunmannedsystems.com/ - manufacturer

Alpha Unmanned Systems is the leading Spanish designer, developer and producer of small Unmanned Aerial Vehicles (UAV) in Spain. The Company is an expert integrator of enhanced technology, Flight Control Systems (FCS) and intelligent payloads for ISTAR missions and non-military tasks.
Alpha Unmanned Systems offers complete solutions including a completely integrated Ground Control Station (GCS), communications, control software and all accessories tailored to the client's requirements

Erle Robotics - http://erlerobotics.com/blog/home-creative/

Erle-Brains are establishing themselves as a leader in the field of lending intelligence to UAVs, we have seen projects including flying defibrillators to heart attack victims, mine clearance using swarm technology and flying gas spectrometers through volcanoes to measure emission impact on global climate. " Says Victor Mayorales. Co Founder Erle Robotics Erle Brain2 brain is multi-platform, embracing the communities of Raspberry Pi, BeagleBoneBlack and Odroid, allowing developers to choose what platform to build over.   Packing USB, Ethernet, I2C, UARTand HDMI, camera, gravity sensor, gyroscope, digital compass, pressure sensor and temperature sensors. This all comes under the most flexible framework for writing robot software, Robot Operating System, (ROS, Indigo). It has support for APM and other Dronecode Foundation tools.

Electronica - http://electronicarc.com/catalogo/ - seller

Hobby Play - http://www.hobbyplay.net/ - seller

Ihobbies - http://www.ihobbies.es/ - seller

Mercado RC - http://www.mercadorc.es/ - seller

RC Innovations - http://rc-innovations.es/ - seller, repairs

RC Tecnic - http://www.rctecnic.com/ - seller

Stock RC - http://www.stockrc.com/ - seller, repairs

UAV Navigation - http://www.uavnavigation.com/ - manufacturer, advanced systems

Wake Engineering - http://www.wake-eng.com/ - manufacturer, advanced systems

Sweden

Bitcraze - https://www.bitcraze.io/ - manufacturer

Hobbytra - http://www.hobbytra.se/ - manufacturer

Intuitive Aerial - http://www.intuitiveaerial.com/ - manufacturer, advanced systems

 The AERIGON Mk II by Intuitive Aerial is an elite cinema drone system for professional cinematographers and broadcast specialists. The AERIGON Mk II's proprietary power distribution system and dual coaxial design with 12 powerful counter-rotating rotors and pre-preg carbon-fiber exoskeleton provides the power, precision and stability to carry the weight of sophisticated camera configurations with pro zoom lenses with full FIZ (Focus, Iris & Zoom) integration with the new DOMINION gimbal/camera controller. The AERIGON Mk II's innovative design absorbs vibrations, conceals and protects cables and electronics from external stress and its detachable arms provide operators the choice between power or endurance, depending on the environment, camera payload or the type of shot requested by the production.

Pitchup - http://www.pitchup.se/ - seller

Switzerland

Aeroscout - http://www.aeroscout.ch/ - manufacturer

Heli Professional - http://www.heli-professional.com/ - manufacturer, seller

Insider Modellbau - http://www.insider-modellbau.ch/ - manufacturer

Pix4D - http://www.pix4d.com/ - drone imaging software

Sensefly - https://www.sensefly.com/home.html - manufacturer, advanced systems

Taiwan

Drones Vision - http://dronesvision.net/ - seller

Falcon Sight Tech - http://www.djitaiwan.com/ - seller

Gaui - http://www.gaui.com.tw/ - manufacturer, advanced systems

Infinity Hobby - http://www.infinity-hobby.com/main/index.php - seller

MKS Servo Tech - http://mks-servo.com.tw/ - servo manufacturer

Savox - http://www.savoxtech.com.tw - electronic components, motor manufacturer

Thunder Tiger - http://www.thundertiger.com/ - manufacturer, seller

Uaver - http://www.uaver.com/ - manufacturer, advanced systems

Thailand

Hobbytime - http://www.phantom4u.com/ - seller

J-Drones - http://store.jdrones.com/default.asp - manufacturer

Turkey

Aselsan - http://www.aselsan.com.tr/en-us/Pages/default.aspx - manufacturer, advanced systems

Baykar Makina - http://baykarmakina.com/ - manufacturer, advanced systems

Eren Hobi - http://www.erenhobi.com - seller

Hobbytime - http://www.hobbytime.com.tr/ - seller

Oyuncak Hobi - http://www.oyuncakhobi.com/ - seller

Pilot Tr - http://www.pilottr.com/ - seller

Promodel - http://promodelhobby.com/ - seller

Sancak Model - http://www.sancakmodel.com/ - seller

TAI - https://www.tai.com.tr/tr - manufacturer, advanced systems

Technomodel - http://www.technomodel.com/ - seller

UAE

Adasi - http://www.adasi.ae/ - manufacturer, advanced systems

Adcom Systems - http://www.adcom-systems.com/ - manufacturer, advanced systems

Ukraine

Fly Technology - http://flytechnology.com.ua/ - seller

Igrotechnika - http://igrotech.com.ua/ - seller

RC Hobby - http://www.rc-hobby.com.ua/ - seller

United Kingdom

Advanced UAV Technology - http://www.auavt.com/ - manufacturer, advanced systems

Birds Eye View Productions - http://www.birdseyeviewpro.co.uk/ - drone services

Blue Bear - http://www.bbsr.co.uk/ - manufacturer, advanced systems

Build Your Own Drone - http://www.buildyourowndrone.co.uk/ - seller

Cobham Antenna Systems -http://www.european-antennas.co.uk/ - drone parts manufacturer

Electric Wingman - https://www.electricwingman.com/ - seller

Flying Gadgets - http://flyinggadgets.com/ - seller

Hobby RC - http://www.hobbyrc.co.uk/ - seller

Hyperflight - http://www.hyperflight.co.uk/ - seller

OrangeRX - http://www.orangerx.uk/ - seller

Oxford Technical Solutions - http://www.oxts.com/ - drone parts manufacturer

Quadcopters - http://www.quadcopters.co.uk/ - seller

Quest UAV - http://www.questuav.com/ - manufacturer, drone services

Radio C - http://www.radioc.co.uk/default.asp - seller

RC Geeks - https://www.rcgeeks.co.uk/ - seller

RC Legends - http://www.rclegends.co.uk/ - seller

Ripmax - http://www.ripmax.com/ - seller, wholesale

UAV Shop - http://www.uavshop.co.uk/ - seller

Unmanned Tech - http://www.unmannedtechshop.co.uk/ - seller

VTOL Technologies - http://www.vtol-technologies.com/ - manufacturer, advanced systems

Westbourne Model - http://www.westbourne-model.co.uk - seller

Zettlex - http://www.zettlex.com/ - manufacturer, advanced systems

312UAV - http://312uav.com/ - seller, drone services

3drobotics - https://3dr.com/ - manufacturer

Action Drone USA - http://www.actiondroneusa.com/ - manufacturer

Adaptive Flight - http://www.adaptiveflight.com/ - flight control systems

Aerial Data Systems - http://www.aerialdatasystems.com/ - drone services

Aerial Media Pros - https://www.aerialmediapros.com/ - seller, consulting, drone services

Aerial Technology - http://www.aerialtechnology.com/ - drone services

Aerouavs - http://www.aerouavs.com/ - drone services, consulting

Aerovel - http://aerovelco.com/ - manufacturer, advanced systems

Agile Sensor Technologies - http://www.agilesensors.com/ - manufacturer

Aircover Integrated Solutions - http://www.aircoversolutions.com/ - manufacturer

Airtronics - http://www.airtronics.net/index.php/ - electronic parts manufacturer

All e RC - http://www.allerc.com/ - seller

Alta Devices - http://www.altadevices.com/ - power systems

Alturn - http://alturn-usa.com - motor manufacturer

Amain Performance Hobbies - http://www.amainhobbies.com/ - seller

Amimon - http://www.amimon.com/ - wireless connections

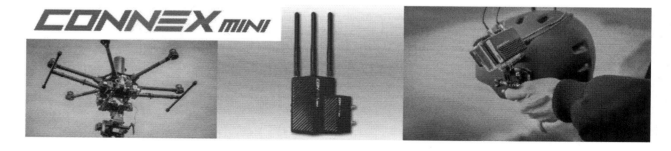

APC Propellers - https://www.apcprop.com/ - propeller manufacturer

Applewhite Aero - http://www.applewhiteaero.com/ - manufacturer, advanced systems

Ares RC - http://www.ares-rc.com/ - seller

Armattan Quads - http://www.armattanquads.com/ - seller

Astro Flight Motors - http://www.astroflight.com/ - motor and electronic parts manufacturer

Atlanta Hobby - http://www.atlantahobby.com - seller

Aurora - http://www.aurora.aero/ - manufacturer, advanced systems

Autel Robotics - https://www.autelrobotics.com/ - manufacturer

Autonomous Nation - http://autonomousnation.com/ - seller

Autonomous Solutions Inc - http://www.asirobots.com/ - manufacturer, advanced systems

Axion Drone Worx - http://www.axiondroneworx.com/ - drone services

Birdseye Aerial Drones - http://birdseyeaerialdrones.com/ - drone services

Blade - http://www.bladehelis.com/ - manufacturer

Buddy RC - http://www.buddyrc.com/ - seller

Carbon by Design - http://www.carbonbydesign.com/ - manufacturer, advanced systems

Century Helicopter Products - http://www.centuryheli.com/ - seller

Cine Drones - http://www.cinedrones.net/ - drone services

Ctrl Me Robotics - http://ctrl.me/ - manufacturer

Cyphy - http://www.cyphyworks.com/ - manufacturer

DIY Quadcopters - http://www.diyquadcopters.com/ - seller

DPI UAV Systems - http://www.dragonflypictures.com/ - manufacturer, advanced systems

Draganfly - http://www.draganfly.com/ - manufacturer

Dragonfly Racing drones - http://www.dragonflyracingdrones.com/ - seller

Dreamqii-Plexidrone - http://www.dreamqii.com/ - manufacturer

Dromida - http://www.dromida.com/ - manufacturer

Drone Fly - http://www.dronefly.com/ - seller

Drone Nerds - http://www.dronenerds.com/# - seller, drone services

Drone World - http://www.drone-world.com/ - seller

Drones Etc - https://www.dronesetc.com/ - seller, repairs

Drones Made Easy - http://www.dronesmadeeasy.com/ - seller

Drones Plus - http://www.dronesplus.com/ - seller

DSLR Pros - http://www.dslrpros.com/ - seller, consulting

Dubro - http://hobby.dubro.com/ - manufacturer, seller

Du-Bro's Tru-Spin Prop Balancer is the most precise balancer on the market. Our specially designed locking cone securely centers and locks props on to the balancing shaft for utmost accuracy. Each cone is carefully inspected on a calibrator to make sure it is running true. The balancing shaft is manufactured out of hardened material and then is centerless ground giving precision balance every time.
The prop balancer's aluminum wheels turn virtually friction free for precision accuracy. The Du-Bro Tru-Spin Prop Balancer will balance model airplane props of all sizes and weights, drone props, boat props, spinners, car wheels, helicopter rotor heads, jet fans, fly wheels, etc.

Eagle Tree Systems - http://www.eagletreesystems.com - autopilot, sensors manufacturer

E-Flite RC - http://www.e-fliterc.com/ - manufacturer

Emax - http://www.emaxmodel.com/ - electronic parts manufacturer

Empire RC - http://www.empirerc.com/ - wholesale

Epson Electronics America Inc. - http://www.eea.epson.com/ - timing and sensing devices

Expert Drones - http://www.expertdrones.com/ - seller, consulting, drone services

Express Drone Parts - http://www.expressdroneparts.com/ - seller

EZ Drone - http://www.ezdrone.com/ - seller, consulting, drone services

Falcon Propellers - http://www.falconpropellers.com/ - propeller manufacturer

Flint Hills - http://www.fhsllc.com/ - manufacturer, advanced systems

Flir - http://www.flir.eu/home/ - advanced imaging and sensing systems

Flite Evolution - http://fliteevolution.com/ - manufacturer

Fly Motion Media - http://www.flymotionmedia.com - drone services

Fly Span - http://www.flyspansystems.com/ - drone services

Get FPV - http://www.getfpv.com/ - parts manufacturer, seller

Gladiator Technologies - http://www.gladiatortechnologies.com/ - drone parts manufacturer

Go Professional Cases - https://goprofessionalcases.com/ - drone and parts cases

Gopro - http://gopro.com/ - cameras

Gothelirc - http://www.gothelirc.com/ - seller

Gowdy Brothers Aerospace - http://www.gowdybrothers.com/ - drone services

Graves RC - http://www.gravesrc.com - seller

Great Planes - http://www.greatplanes.com/ - manufacturer, seller

Hacker Motor Distribution Co. - http://www.hackermotorusa.com - motor manufacturer

Helidirect - http://www.helidirect.com/ - seller, repairs

Helimax - http://www.helimaxrc.com/ - manufacturer

Hexoplus - http://www.hexoplus.com/ - manufacturer

Hobbico - http://www.hobbico.com- manufacturer, seller

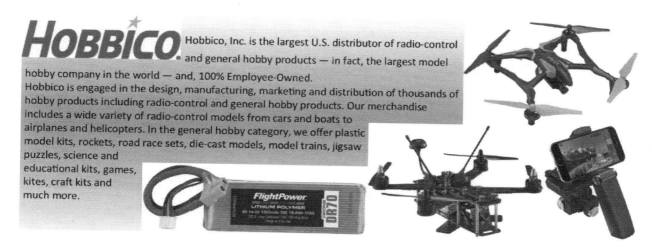

Hobbico, Inc. is the largest U.S. distributor of radio-control and general hobby products — in fact, the largest model hobby company in the world — and, 100% Employee-Owned.
Hobbico is engaged in the design, manufacturing, marketing and distribution of thousands of hobby products including radio-control and general hobby products. Our merchandise includes a wide variety of radio-control models from cars and boats to airplanes and helicopters. In the general hobby category, we offer plastic model kits, rockets, road race sets, die-cast models, model trains, jigsaw puzzles, science and educational kits, games, kites, craft kits and much more.

Hobby Express - http://www.hobbyexpress.com/ - seller

Hobbyking - http://www.hobbyking.com/ - manufacturer, seller

Hobbyuav - http://hobbyuav.com/ - manufacturer

Hobbyzone - https://www.hobbyzone.com/ - seller

Horizon Hobby - http://www.horizonhobby.com/ - seller

Hylio - http://www.hyl.io/ - manufacturer

Iftron Technologies - http://www.iftrontech.com/ - seller

Innova Drone - http://www.inovadrone.com/ - manufacturer

Innovative UAS - https://www.innovativeuas.com/ - seller, consulting, drone services

Intelligent UAS - http://1uas.com/ - seller, repairs

Iron Ridge - http://www.ironridgeuas.com/ - manufacturer

Leptron - http://www.leptron.com/ - manufacturer, advanced systems

Lipomall - http://www.lipomall.com/ - power systems seller

LRP America - http://www.lrp-americastore.com/ - manufacturer, seller

Madlab Industries - http://www.madlabindustries.com/ - manufacturer

Magic Sky - http://www.magicskyusa.com/ - seller

Maniacs Hobby - http://www.maniacshobby.com/ - seller, repairs

Marcus UAV - http://www.marcusuav.com/ - manufacturer, advanced systems

Mars Parachutes - http://www.marsparachutes.com/ - drone parachutes

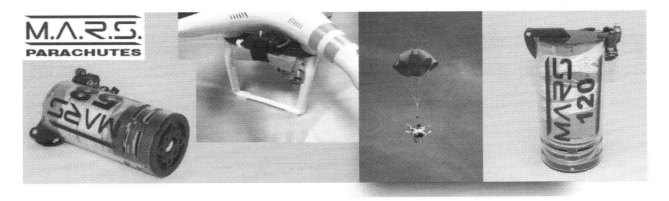

Martin UAV - http://martinuav.com/ - manufacturer, advanced systems

Maverick Drone Systems - http://www.maverickdrone.com/ - seller, drone services, consulting

Maxbotix - http://www.maxbotix.com/ - ultrasonic sensor manufacturer

Maxx Products International - http://www.maxxprod.com/ - parts manufacturer

Memsic Inc. - http://www.memsic.com/ - sensors manufacturer

Milsource - http://militaryethernet.com/ - seller

MKS Servo USA - http://www.mksservosusa.com/ - servo manufacturer

Moog Crossbow - http://www.moog-crossbow.com/ - advanced inertial sensor systems

MPI Motors - http://www.mpimotors.com - motor manufacturer

Multi Rotor Superstore - https://www.multirotorsuperstore.com/ - seller

Multicopter Warehouse - http://www.multicopterwarehouse.com/ - seller

Natural Drones - http://www.naturaldrones.com/ - manufacturer

 Naturaldrones design, develop, build and operate high end professional sUAS systems with a specific attention to safety requirements. We propose systems equipped with Optical, Multispectral and Thermal sensors, ready to be used in the fields of Critical Infrastructure Monitoring, Environmental Monitoring, Geomatics, Construction Monitoring, Precision Agriculture.

New Creations RC - https://www.newcreations-rc.com/ - seller

Nitro Planes - http://www.nitroplanes.com/ - seller

No Limit UAV - http://store.nolimituav.com/ - seller

Nuwaves Engineering - http://www.nuwaves.com/ - parts manufacturer, advanced systems

Omnetics - http://www.omnetics.com/ - connector manufacturer

Onagofly - http://www.onagofly.com/ - manufacturer

Openpilot - http://www.openpilot.org/ - open source autopilot

Peau Productions, Inc. -http://www.peauproductions.com/ - camera manufacturer

 Peau Productions provides lenses and filters for a large catalog of small cameras such as the GoPro Hero, Sony Action Cameras, Xiaomi Yi, and DJI Phantom and X3. The majority of our lenses concentrate on low distortion, non-fisheye optics in various focal lengths and high megapixel ratings.

MAPIR cameras provides light-weight, affordable and easy to use surveying cameras for industries such as agriculture and construction. Each model line comes in various spectral options allowing a user to use as few or as many as they require for their survey. www.mapir.camera

Pelican Drones - http://www.pelicandrones.com/ - seller, drone services

Polar Pro Filters - https://www.polarprofilters.com/ - camera accessories

PolarPro is the leading manufacture of drone accessories with a specialty in premium, ultra-lightweight filter solutions. We offer products for the Phantom 3, Phantom 4, Inspire 1 (X3), Inspire Pro (X5), 3DR Solo, and the Yuneec Typhoon. We also offer a wide range of filters for drones using the GoPro camera, as well as an extensive line of GoPro accessories. PolarPro continues to innovate and develop products for helping users capture the highest quality video and pictures possible.
Some of our most popular products are:
-Phantom 3 & Phantom 4 Filters
-Phantom 3 Landing Gear
-3DR Solo Prop Guards
-Phantom 3 Light Kit
-3DR Solo Light Kit
-GoPro Frame2.0 Filters
-Phantom 3 Gimbal Guard

Power Drone Sales - http://www.powerdronesales.com/ - seller, consulting, repairs

Power Hobby - http://www.powerhobby.com/ - seller

Quadrocopter.com - http://www.quadrocopter.com/ - seller, drone services

QuantumRC - http://www.quanum-rc.com/index.php - manufacturer, seller

RC Depot - https://www.rcdepothobbies.com/ - seller

RC Planet - http://www.rcplanet.com/ - seller

RC Rotors and Aerial Media - http://shop.rotorsandaerialmedia.com/ - manufacturer, advanced systems

RC Superstore - http://www.rcsuperstore.com/ - seller

Reactel Inc. - http://www.reactel.com/ - parts manufacturer, advanced systems

ReadymadeRC - http://www.readymaderc.com/store/ - seller

Realflight - http://www.realflight.com - drone flight simulations

Red Rocket Hobbies - https://www.redrockethobbies.com/ - seller

Robart - www.robart.com - parts manufacturer, seller

Rotor Logic - http://www.rotorlogic.com/ - seller

Run Cam - http://www.runcam.com/ - camera manufacturer

Sadler Aircraft Corporation - http://www.sadlerair.com/ - maunfacturer

Sagetech Corporation - http://sagetechcorp.com/index.html - transponder manufacturer

SBG Systems North America - http://www.sbg-systems.com/ - inertial sensors manufacturer

Scion UAS - http://www.scionuas.com/ - manufacturer, advanced systems

Scorpion Drones - http://www.scorpiondrones.com/ - seller

Seeed - http://www.seeedstudio.com/index.html - electronic components manufacturer

Shot Over - http://shotover.com/ - manufacturer

Silent Falcon - http://www.silentfalconuas.com/ - manufacturer

Small Parts CNC - http://www.smallpartscnc.com - parts manufacturer

Smp Robotics - http://www.smprobotics.com - UGV manufacturer

SMP Robotics develops and provides light unmanned ground vehicles UGV designed for all-season driving in designated areas. over S5 performs multiple trips over mapped routes, unattended. Its autonomous motion control and navigation system is guided by machine vision, performing pinpoint video data analysis, which allows the vehicle to select and navigate routes, avoid obstacles, stay on course and repeat the job cycle multiple times.

Sky Pirate - http://www.skypirate.us/ - manufacturer, seller, drone services

Skygen Aviation - http://www.skygenaviation.com/ - manufacturer

Skygrower - http://store.skygrower.com/ - seller

Skyspecs - http://www.skyspecs.com/ - manufacturer

 SkySpecs is a UAV technology company located in Ann Arbor, MI. SkySpecs delivers a fully autonomous drone inspection that saves company's time, resources and money, and delivers the highest quality data with the least amount of interference or disruption. We work with the wind energy industry to enable fast, reliable wind turbine inspections. Our team of expert robotic and computer science engineers are developing the technology that will enable a full spectrum of industry services in 2017.

Spectracom - http://spectracom.com/ - manufacturer, advanced systems

Sure RC - http://www.surerc.com/ - seller

Tattu Battery - http://www.genstattu.com/ - seller

Think RC - http://www.thinkrc.com/ - seller

Tor Robotics - http://torrobotics.com/ - manufacturer

Tower Hobbies - http://www.towerhobbies.com/ - seller

Traxxas - http://www.traxxas.com/ - manufacturer

Aton™ is your personal video assistant that captures stunning aerial footage in a way that is easy and fun, without the need for any specialized skill or experience. Aton is sporty and smart, practically flying itself with auto take off and simple, intuitive directional control. You simply direct Aton towards your subject and Aton becomes the steady hand in the sky that captures the moment from exciting new vantage points. Press a button, and Aton quickly returns to home and lands automatically.

Alias is a clean-sheet design that focuses on the performance and design elements that make flying, and even learning to fly, fun. The first mission objective was durability. Alias is built around a unique molded-composite frame that is combined with clever high-tech construction to make it extremely light and virtually indestructible. Your flying fun goes on and on without worrying about breakage from crash damage.

Triad RF - http://www.triadrf.com/ - amplifier systems

Troy Built Models - http://www.troybuiltmodels.com/ - seller

Turnigy - http://www.turnigy.com/ - electronic components, power systems manufacturer, seller

UAI International - http://www.uaiinternational.com/ - drone services

UAV Factory - http://www.uavfactory.com/ - manufacturer, advanced systems

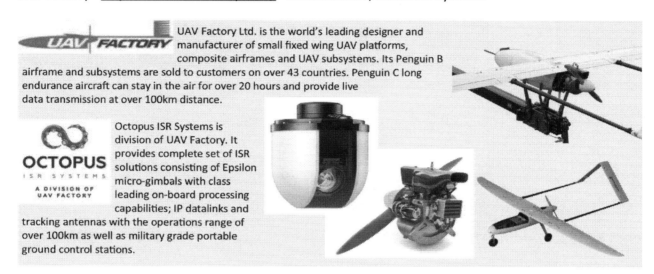

UAV Factory Ltd. is the world's leading designer and manufacturer of small fixed wing UAV platforms, composite airframes and UAV subsystems. Its Penguin B airframe and subsystems are sold to customers on over 43 countries. Penguin C long endurance aircraft can stay in the air for over 20 hours and provide live data transmission at over 100km distance.

Octopus ISR Systems is division of UAV Factory. It provides complete set of ISR solutions consisting of Epsilon micro-gimbals with class leading on-board processing capabilities; IP datalinks and tracking antennas with the operations range of over 100km as well as military grade portable ground control stations.

UAV Products - http://www.uavproducts.com/ - seller

UAV Solutions - http://www.uavsolutions.com/ - manufacturer, advanced systems

Phoenix ACE LE UAS is delivered system ready with a backpack, Windows tablet and joystick for operation, communication radios and a choice of payload. Modular payload options include an electro-optic / FLIR Vue infrared camera, a Sony 10x zoom camera, a Canon S100 camera and a MicaSense Sequoia multispectral camera.

With its foldable booms, the Phoenix 30 quad rotor is easy to transport and easy to store. The system endurance is 30 minutes and this air vehicle can fly in up to an inch-an-hour of rain. The UAS is fully autonomous with multiple user modes.

Made from 7075 grade aluminum, EPP foam and composite materials, the Talon 120LE is a hand-launched, belly recovery vehicle with endurance up to 2.5 hours. The system is fully autonomous with multiple user modes.

UAVionix - http://www.uavionix.com/ - transmitter, receiver, navigation manufacturer

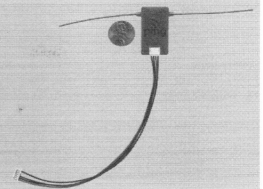

At present, there is no universal way for both unmanned and manned aircraft operators to visualize the airspace. Manned aircraft have a system in place to detect other manned aircraft, but unmanned aircraft systems (UAS) are not broadly integrated into that system. uAvionix's ADS-B receivers and transponders utilize the existing FAA communication infrastructure used by manned airplanes, providing drone pilots the same degree of situational awareness as pilots and controllers of manned aircraft. In other words, a drone equipped with the Ping ADS-B transponder will not only sense-and-avoid other drones, but also manned aircraft. This has huge potential to solve many of the existing safety and policy challenges.

UAVX - http://www.uavx.com/ - flight control systems

Ulti-Mate Connector Inc. - http://www.ultimateconnector.com/ - connector manufacturer

Unmanned Systems Source - http://www.unmannedsystemssource.com/ - seller

Vertical Partners West - http://www.vpwllc.com/ - manufacturer

Video Aerial Systems - http://videoaerialsystems.com/ - manufacturer

Vision Aerial - http://www.visionaerial.com/ - manufacturer

Vision Aerial is a US-based manufacturer of professional drone platforms and payloads. Its drones are flying in over 15 countries on three continents. The company has received numerous awards including Best of What's New from Popular Science. All unique components are produced in Bozeman, Montana where the entirety of the company's operations are located.

Xiro - http://www.xirodroneusa.com/ - manufacturer

Xoar International - http://www.xoarintl.com/ - propeller manufacturer

Xproheli - http://www.xproheli.com/ - drone services

Drone Events

Age of Drones - http://ageofdrones-expo.com/

AUVSI - http://www.auvsi.org/home

Commercial UAV Expo - http://www.expouav.com/

Drone Racing League - http://thedroneracingleague.com/

Drone Wars - http://dronewarsfpv.com/

Drone World Expo - http://www.droneworldexpo.com/

Drone Worlds - http://droneworlds.com/

Europa Drone - http://www.europadrone-event.com/

Interdrone - http://www.interdrone.com/

International Drone - http://www.internationaldroneday.com/

International Drone Expo (IDE). http://internationaldroneexpo.com/

Skytech events - http://www.skytechevent.com/

Texas UAS Summit - https://www.texasuassummit.com/

UK Drone Show - http://www.ukdroneshow.com/

Unmanned Systems Technology - http://www.unmannedsystemstechnology.com/events/

Drone Discussion Forums

http://diydrones.com/

http://www.roboticmagazine.com/forums

http://www.rcgroups.com/forums/index.php

http://www.hobbyking.com/hobbyking/forum/

http://www.multirotorforums.com/

http://www.droneflyers.com/talk/

http://www.ardupilot.com/forum/

http://www.multiwii.com/forum/

Drone News Sources

UAS Magazine - http://www.uasmagazine.com/

That Drone Show - http://www.thatdroneshow.com/

Beginner Flyer - http://beginnerflyer.com/

Unmanned Systems Technology - http://www.unmannedsystemstechnology.com/

Robotic Magazine- http://www.roboticmagazine.com/

The UAV Digest - http://www.theuavdigest.com/

Quadcopterflyers - http://www.quadcopterflyers.com/

Rotor Drone Magazine - http://www.rotordronemag.com/

SUAS News - http://www.suasnews.com/

The Drone Files - http://www.thedronefiles.net/

Drone Flyers - http://www.droneflyers.com/

Droneblog - http://droneblog.com/

Naval Drones - http://www.navaldrones.com/

Dronestagram - http://www.dronestagr.am/

Drone Racing World - http://droneracingworld.com/

Drone Overdose - http://www.droneoverdose.com/

Large Online Retailers which also sell hobby grade drones and parts

aliexpress.com	amazon.com	banggood.com
bestbuy.com	gearbest.com	geekbuying.com
target.com	tinydeal.com	tmart.com

Future of drones

- Increased power efficiency, battery strength, reduced noise, increased operation time, increased speed, altitude and range, increased autonomous abilities. These will also increase areas of use and carrying capacity. Many future possibilities greatly depend on battery power.

- Size of drones will decrease which means they will have less kinetic energy on impact in case of a fall and therefore it may ease regulations

- Manual piloting a drone will be history, unless specifically desired by the user.

- Delivery drones may be common place. Drones will be able to carry more weight, and drone parachutes may be a mandatory feature. There may be drone charging poles at every certain distance, where, drones automatically attach and charge themselves.

- We may have our own drones, similar to having cars, to pickup any items we need, when we want, not only from shopping but also anything else too. With increases in flight time, we may also see our personal drones following us to constantly from a close distance to record our activities, such as when we go mountain climbing, skiing, sailing...

- Data gathering about our surroundings will greatly improve, even in real time

- Construction: Attachment of arms, or tools, which will open a whole new world of possibilities, such as in construction. There are countless tasks in construction, where materials must be elevated or workers must work above ground level, or, as in most cases, both of these taking place at the same time. Drones can carry things up and down, and skilled craftsmen on the ground can direct the drone arms through FPV, or some tasks may take place autonomously with the improvements in robotics. Going even further to the future, the autonomous abilities of robots will improve to the point that they may not need humans, to direct its arms when doing a certain construction installation. So all they may need is a construction plan or sketch uploaded to them, and then they will do the work autonomously, also by communicating to each other. Also remember that one of the biggest challenges in robotics is to make a robot move itself on land in an undefined environment. With the use of drones that fly, this challenge is eliminated altogether too. All this will increase construction speeds and reduce costs. Drones already started to be used such as visual inspection of elevated surfaces where normally a crane or scaffolding would have been required to access.

- Law enforcement applications will increase due to increased battery power and autonomous abilities, such as tracking of suspects, surveillance, remote presence

- Military Applications:

Air minefield, or swarm, such as a network of small drones in big numbers swarming around enemy airplanes, helicopters or rockets to force them to change direction or interfere with them. A similar application of this was already tested on the sea by US Navy in 2014 which involved small autonomous ships in big numbers swarming enemy ships. Keep in mind that the drones of the future will be able to stay in the air much longer time, for much cheaper price. They can charge themselves as needed by returning to charging stations now and then. A big network of totally autonomous drones, spacing themselves at a certain distance from each other and constantly communicating location of each other by using swarm robotics concepts and location and projected path of enemy planes can provide an air defense mechanism, by communicating the projected path to drones ahead. With small guns or bombs attached, the effectiveness and coverage radius will increase, or they can even use electromagnetic waves to jam the electronic devices of the enemy aircraft that passes by.

Manned warplanes may be eliminated greatly and be replaced by drone airplanes. The battle of two countries air forces may only mean their drone jets are fighting.

Land drones, such as drones operating very close to land surface, with light weapons or bombs attached, to act as land soldiers, for more precise operations than of today's armed drones, which shoot from distance. The advantage of these against a walking unit would be that they will be much faster than a walking unit, plus they will not need to tackle the challenges of walking on irregular surfaces. They will be very close to surface, but still take advantages of flying. The movement and weapon operation of drones can be autonomous or remote controlled as determined on a case by case basis.

Future Impacts on economy: As happened always since industrial revolution, each automation creates higher quality and more information based jobs even if it eliminates some old professions. Just like the industrial revolution replaced many people working in the farms with machines, who started doing something else. For instance in this case, due to the benefits mentioned above, we will need less cars and less infrastructure as roads, less auto mechanics for cars, less drivers, but more people will be needed to create and manage the software and the process and the new businesses with drones which is a more information based job. Autonomous driving cars will also be common place when drones will be in our lives in this way, and delivery and transportation by autonomous cars will also be the main method of delivery but regardless of that, drones will still share the increased workload. In the future the volume of goods delivery will increase, even if we will have 3d printers at every house or office and producing more and more variety of items. What we had written for autonomous cars almost 3 years ago, in an article called "An in depth analysis of autonomous cars" also applies to drones:

"In the end, this will just be viewed as any other simple automation that happened in our lives, like many other things that we take for granted today but is automated already, such as robots making cars, or software checking credit card fraud automatically. Finally, in 2050, at a time when the manual driving will for so long be history, our children will look at the old movies of today, and our manual driving of cars will seem the same way we perceive horse carriages today."

In 2050, where we will see many, many drones everywhere and everyday, doing much more tasks with more abilities, flying much longer and carrying heavier loads, we will perceive our droneless skies of today as the old times.

Made in the USA
Middletown, DE
11 June 2017